SpringerBriefs in Fire

Series Editor

James A. Milke

For further volumes:
http://www.springer.com/series/10476

Joshua Dinaburg · Daniel T. Gottuk

Fire Detection in Warehouse Facilities

 Springer

Joshua Dinaburg
Daniel T. Gottuk
Hughes Associates, Inc.
Baltimore, MD
USA

ISSN 2193-6595   ISSN 2193-6609   (electronic)
ISBN 978-1-4614-8114-0   ISBN 978-1-4614-8115-7   (eBook)
DOI 10.1007/978-1-4614-8115-7
Springer New York Heidelberg Dordrecht London

Library of Congress Control Number: 2013943571

Printed on acid-free paper

Springer is part of Springer Science+Business Media (www.springer.com)

Foreword

Automatic sprinklers are the traditional fire protection system in warehouses; their major purpose is to control and suppress fire. However, with an increase in warehouse sizes, limitations in water supply and changes in firefighting strategies by fire departments, the effectiveness of current sprinkler protection strategies for warehouses has been questioned. In 2009 and 2010, the Fire Protection Research Foundation conducted two workshops to address fire safety concerns in modern warehouses, which involved warehouse users, insurance companies, fire engineering firms, researchers, fire protection system manufacturers, and code and standard writers[1,2]. At these workshops, the application of fire detection for early fire warning, fire location identification and monitoring in warehouses was explored, as were the potential benefits of quicker response of suppression systems, reducing water supply requirements, and minimizing the involvement of fire departments. However, currently there is little research available or guidance on the utilization of fire detection technologies of various types in warehouse environments.

Accordingly, the Fire Protection Research Foundation initiated this project with an overall goal to provide technical information for the development of guidelines and standards for applying fire detection technologies for modern warehouse protection. The objective of this first Phase program was to assess the potential role for detection systems in this environment and develop a proposed research plan to implement the goal.

The Foundation acknowledged the contributions of the following individuals and organizations to this project:

Project Technical Panel

Bob Elliott, FM Approvals
Rich Gallagher, Zurich Insurance
Dave Icove, University of Tennessee
Jonathan Levin, Liberty Mutual Property
Karen Rebman, Target
Nicholas Williams, C&S Wholesale Grocers
Steve Wolin, Code Consultants, Inc.

Project Sponsors

AP Sensing
Axonx
Fire Fighting Enterprises
SimplexGrinnell
Xtralis

Contents

Fire Detection in Warehouse Facilities 1

1 Background 1

2 Objective 1

3 Approach 1

4 Literature Review 3

   4.1 Existing Warehouse Characteristics and Fire
Protection Strategies. 3

   4.2 Fire Loss Incidents. 8

   4.3 Applicable Detection Systems 14

   4.4 Warehouse Detection Research 18

   4.5 Literature Review Conclusions 24

5 Fire Hazard Analysis 25

   5.1 Characteristic Warehouse Designs 26

   5.2 Characteristic Warehouse Fires 29

   5.3 Warehouse Fire Analysis 46

   5.4 Additional Applications of Detection Equipment 51

   5.5 Fire Hazard Analysis Summary and Conclusions 52

6 Research Plan 53

References 57

Bibliography of Additional Sources 61

Fire Detection in Warehouse Facilities

1 Background

Automatic sprinklers systems are the primary fire protection system for use in warehouse and storage facilities. The effectiveness of this strategy has recently come into question at the National Fire Protection Research Foundation Workshops conducted in 2009 and 2010 [1, 2]. These concerns arose due to the increasing challenges presented by modern warehouse facilities, including increased storage heights and areas, the use of automated storage retrieval systems (ASRS), limitations in available water supplies, and changes in firefighting strategies. The application of fire detection devices used to provide early warning and notification of incipient warehouse fire events is being considered as a component of modern warehouse fire protection.

2 Objective

The purpose of this project is to provide technical information to aid in the development of guidelines and standards for the use of fire detection technologies for modern warehouse fire protection. The goal of this phase one project is to assess the potential role for fire detection in modern warehouse facilities and to develop a proposed research plan to implement the overall project goal.

3 Approach

The phase one project was broken into three distinct tasks. The first task consisted of a literature review focused upon various aspects of the warehouse fire detection issue. These aspects included:

1. A characterization of existing warehouse facilities, both within the United States and abroad. This included a review of the number and size of warehouse

J. Dinaburg and D. T. Gottuk, *Fire Detection in Warehouse Facilities*,
SpringerBriefs in Fire, DOI: 10.1007/978-1-4614-8115-7_1,

facilities, as well as the general design characteristics and features, both voluntary and compulsory. Some additional discussion was included regarding the general fire department response and actions in various jurisdictions for warehouse fire incidents.

2. A review of modern fire loss incidents in warehouses, including both statistical and narrative components. The review focused upon the typical loss scenarios, the ignition, flame spread, and commodity types, the presence and role of fire protection features, and the actions of the fire department.
3. The types of fire detection equipment that may be applicable to warehouse installations.
4. Existing research relevant to the application and performance of fire detection systems in warehouses.

The second task consisted of development of a fire hazard assessment for modern warehouse facilities. The assessment was intended to provide insights into the potential impact of early warning detection and notification with a goal of reducing the destruction of property during warehouse fire incidents. The fire hazard analysis focuses upon four factors related to the severity and impact of warehouse fire incidents. These aspects include:

1. Development of characteristic warehouse building designs, including building sizes, storage configurations, and suppression systems. The proposed warehouse designs are intended to encompass a variety of designs that are compliant with NFPA 13 requirements [3].
2. Development of characteristic fire incidents based upon statistically significant ignition scenarios and locations. Mathematic correlations have been used to estimate potential fire growth, smoke production rates, and likely sprinkler activation times of the fires within the design warehouses. The potential extent and growth rate of each fire has been considered both with and without operating and/or effective sprinkler systems present.
3. Description of the general fire fighting assessments and tactics used for fighting warehouse fires is provided. The expected fire department response times occurring after discovery of a fire event are also reviewed and compared with the calculated fire growth calculations to estimate the condition of the warehouse at the time fire combat is ready to begin.
4. Evaluation of the potential impact of detection technologies with regard to the fire department response time after ignition and the severity of the fire at that time. An assessment has been made to determine the potential ability to provide earlier warning to reduce the total response time and reduce the severity of the fire at the beginning of combat operations. Individual detection methods and devices are not reviewed for this analysis, but rather a general time window available to any detection technologies has been considered.

The third and final phase of the overall project was to identify information gaps and recommend future work to aid in the development of guidelines and standards for the use of detection technologies in warehouse fire protection design.

4 Literature Review

4.1 Existing Warehouse Characteristics and Fire Protection Strategies

Warehouse properties include a vast array of facilities intended for the storage of various commodities. Individual warehouses may vary greatly depending upon the type and quantities of materials stored, the intermixing of multiple stored commodities, the level of automation used for storage and retrieval, the height and method of stacking commodities, the existing fire protection and life safety features, and the size, shape, and construction of the buildings themselves. With such variation, it is not a simple task to classify a "typical" warehouse for the purposes of fire protection design. Depending upon the jurisdiction, or even global region or nation under consideration, mandated design principles are enforced and can influence the overall implementation of storage method and building design, fire protection devices, and firefighting tactics.

Important factors in warehouse design for fire protection implementation include the building area and height, the storage methods, the stored commodity types, existing fire protection equipment, as well as local rules and regulations. In addition, the general fire fighting tactics used in the event of a warehouse fire should be considered. Where information was available, the general warehouse design characteristics, including fire protection concerns, regulations, and tactics have been discussed.

4.1.1 United States

According to the Unites States Department of Energy—Energy Information Administration (US DOE EIA), the United States has been estimated to contain approximately 600,000 warehouse facilities as of 1999. The average warehouse contains a storage area of 1,600 m^2 (17,400 ft^2). Among the 600,000 EIA recognized warehouse properties, many are sized well below the average floor area, with 44 % sized between 93–465 m^2 (1,001–5,000 ft^2) and 70 % less than 930 m^2 (10,000 ft^2) in storage area. The overall average floor area is elevated due to the increasing number of extremely large warehouses, with over 50,000 warehouses reported to be larger than 4,645 m^2 (50,000 ft^2) [4]. It is within these large warehouses, where fire loads could become immense, that the greatest potential for large, destructive fires exists.

As the desire for greater economy is emphasized, warehouse facilities stress the need to increase storage heights and reduce open floor space. It is not uncommon for a warehouse space to store commodities at heights greater than 9–12 m (30–40 ft) over an area as large as 10 football fields, or 54,000 m^2 (576,000 ft^2). When automatic storage and retrieval systems (ASRS) are used, storage may exceed heights of 30 m (100 ft) [5].

ASRS systems consist of motorized, electronic, computer controlled equipment used to place and retrieve stored materials from within racks or other distributed storage systems. The use of motorized carts allows material to be stacked higher and with narrower aisles than for manned warehouse spaces. While this does provide an increased level of efficiency, both in retrieval times and the use of space, the ASRS does introduce some unique threats for fire protection design. The additional fire risk of utilizing ASRS systems include:

- The electronics required present potential ignition hazards.
- The cart may obstruct detection or suppression equipment from targeting the seat of a fire.
- The retrieval system and narrow aisles increase difficulty and hazard for personnel to manually respond to a fire incident. Increased storage heights with ASRS further limit ability to provide manual response.
- The increased storage density and heights increase the potential size and growth rate of fires.
- The increased storage density limits the ability of suppression systems to penetrate.
- Fewer employees on-site decrease chances of personnel detecting a fire (this places greater emphasis on automatic fire detection systems).

The increased efficiency and economy of ASRS systems have resulted in more widespread use and the necessities for additional fire protection features.

Warehouses generally contain commodities classified by *NFPA 1—Fire Code* [6], Sect. 34.2.5 (or International Fire Code (IFC) Chap. 23 [7]). Commodities are categorized according to the material types and combustion properties into various Classes, where Class I is the least threatening and Class IV is has the most associated risks. The distinctions between Classes are determined based upon the amounts, arrangements, and types of combustible materials including the packing materials and pallets. Classified materials may include wood, paper, cloth, and plastics. Plastics are grouped into Group A, B, and C materials depending upon the heat of combustion, with A being the most hazardous. The inclusion of plastics within the commodities and packaging generally act to increase the hazard classification. The various commodity types are used to determine the fire hazards and required fire protection design criteria.

The general practice in the United States is to utilize automatic sprinkler suppression systems for the protection of warehouse storage properties when required for the storage configurations and commodity types by applicable building or fire codes and insurance requirements. The design of sprinkler systems for storage spaces is determined by the criteria of *NFPA 13—Standard for Installation of Sprinkler Systems* [3]. Additional fire protection requirements may include the use of fire dampers, smoke partitions, fire rated barriers, smoke and heat vents, and fire proofing of building columns and roofs.

A typical device used for detection of fires in these facilities is a water flow switch, used to alert occupants and occasionally central stations or fire departments when a sprinkler head has activated. Depending upon the ignition location and the height of installed sprinkler heads, detection and notification of a fire by sprinkler activation may not take place until fires have developed into significant threats.

For many modern warehouse properties, the provisions of the NFPA codes are not sufficient to provide guidance for adequate sprinkler coverage or design. For example, storage of exposed expanded plastic on racks is not currently addressed by NFPA 13 requirements. Some of the unique challenges presented by increasing storage heights and building heights, storage arrangements, or commodities are not directly addressed by standards such as NFPA 13. Often, a performance based design may be necessary, but modeling is extremely limited to predict the flame growth of large racked storage and even more limited to predict the ability of sprinklers to suppress such flames. The variables affecting the design of a sprinkler system include commodity classification, packaging materials, storage configurations, aisle widths, storage heights, pile stability, sprinkler clearances, as well as sprinkler types [8].

The use of various sprinkler technologies for warehouse properties has evolved over the past several decades. Investigations into sprinkler performance for tall storage indicated increased suppression and control performance for large orifice and droplet sizes due to the ability of the water to penetrate to the seat of the fire. Such research led to the development and implementation of control mode specific application (CMSA) and Early Suppression Fast Response (ESFR) sprinklers for the protection of higher storage without the need for in-rack sprinklers [9]. As storage heights and floor areas continue to increase, such water application methods may no longer be sufficient for the control and suppression of warehouse fires and more sophisticated options may be necessary.

Fire fighting tactics and strategies for warehouse properties are often dictated by the size and extent of the fires upon arrival, but are often generally defensive in nature. Due to the potential for rapid flame spread and limited notification procedures, warehouse fires are often significant thermal threats at the time of fire department arrival. This is also exacerbated by warehouses often being located in fairly remote locations due to cheap and available land for large facilities. If firefighters arrive at a facility with a high level of smoke or well developed fire, they may not know if the fire is controlled by suppression systems, if the building steel is structurally sound, whether unstable stock could collapse upon entry, or when to ventilate. Generally, the fire departments are left making a safe assessment not to risk entry [5].

4.1.2 Europe

The issue of increasing warehouse size has also been investigated by fire protection researchers in Europe. Recent studies conducted in the United Kingdom (UK) by the Building Research Establishment (BRE) determined that the average warehouse size had increased from 400 m^2 (4,300 ft^2) in the 1970s to 1,700 m^2 (18,300 ft^2) for existing facilities in 2002. The study also revealed that planned construction of new warehouse properties at that time averaged 7,100 m^2 (76,400 ft^2). It was also reported that warehouse properties of 25,000–40,000 m^2 (270,000–430,000 ft^2) were not uncommon. Typical ceiling heights in such facilities are 11.5–12 m (38–40 ft) with commodities stored at heights of 10.5 m (35 ft) [10].

Utilizing an estimate provided in British Standard PD 7974-7 that 3.3 × 10^{-5} fires occur per square meter of warehouse space per year and the estimated 2,500 annual warehouse fire incidents, the BRE study determined that there are approximately 30,000 warehouse properties in the UK [10]. Among these properties, it has been determined that large single story warehouses over 2000 m^2 (21,500 ft^2) with mezzanines, galleries, high bay racking, and ASRS systems are posing the greatest current threat. According to the Fire Safety Advisory Board Chair, Pamela Castle, the risks of modern warehouses in the UK are increasing due to the use of remote controlled storage retrieval, insulating core panels, plasticized goods and storage, and factory built construction [11]. Such methods increase efficiency but simultaneously increase potential fire loads and hazards, and often result in faster fire spreads. In addition, the replacement of wooden with plastic pallets has introduced additional hazards for flame spread and noxious emissions.

Fire protection design of these facilities is governed by the Fire Precautions (Workplace) Regulations 1997 [12]. This document describes the requirements for performing a fire risk assessment and prescriptions for obtaining a Fire Certificate. In addition, the Regulatory Reform (Fire Safety) Order 2005 provides prescriptive elements for design, including the establishment of a responsible manager and the elements required in the design [13]. In general, however, the specific design elements are left to the discretion of the manager, including automatic fire fighting and detection equipment, fire department access and contacts, and overall fire safety planning.

The BRE stresses that UK regulations are focused upon the health and safety of persons in buildings, and not upon reducing the costs of property damages. It is generally believed that this is a responsibility of the insurance industry, and should not be a primary concern for compulsory building regulations. The use of sprinklers in warehouse properties is not expected to reduce the limited number of injuries and deaths occurring on these properties, and for this reason, no national statute mandates the use of sprinklers in such properties [10]. In addition to sprinkler requirements, the UK also does not issue statutory requirements regarding property protection or fire department access [11].

Some properties do still install sprinklers, but it is more closely related to insurance requirements and the length of rental contracts based upon receiving a financial return on investment. Some jurisdictions do enforce local prescriptive requirements for sprinkler use in warehouse properties. Through interviews and surveys, it was estimated that 20–50 % of warehouse owners would voluntarily sprinkler building [10].

Some effort has been enacted to require the use of sprinklers in extremely large storage properties. A research project was enacted by the BRE to determine if mandating sprinklers in the largest of warehouse, specifically those over 440,000 m^2 (4.7 million ft^2) would be cost appropriate. Cost-benefit analysis assessing the risks and costs of warehouse fires determined with 90 % confidence that sprinkler installation would be financially beneficial in warehouses larger than 85,000 m^2 (915,000 ft^2) [10]. Research has indicated that fire costs are reduced by 80 % in sprinklered buildings and that a sprinkler system utilizes 15 times less water than that used by fire department hoses [14].

Additional European countries do mandate the installation of sprinklers in storage properties exceeding a certain size, generally much smaller than the proposed BRE limitation. Sprinklers are required in buildings of area exceeding [10]:

- 5,000 m^2 (54,000 ft^2) for normal hazard and 2,000 m^2 (21,500 ft^2) for higher hazard in Denmark
- 3,000 m^2 (32,300 ft^2) in France
- 1,200 m^2 (13,000 ft^2) for normal hazard and 400 m^2 (4,300 ft^2) for higher hazard in Germany
- 1,000 m^2 (10,800 ft^2) in Netherlands
- 1,800 m^2 (19,400 ft^2) for normal hazard and 1,200 m^2 (13,000 ft^2) for higher hazard in Norway
- 2,000 m^2 (21,500 ft^2) for normal hazard and 1,000 m^2 (10,700 ft^2) for higher hazard in Spain.

In the event of large warehouse fires, the fire departments in the UK often do not attempt to enter the buildings unless immediate rescue operations are necessary. In addition, water application is generally limited from the outside until the buildings have collapsed [11]. Similar to the regulations, the fire department response to large warehouse fires is primarily focused upon life safety with little regard to the protection of property. Current research is being conducted by the International Technical Committee for the Prevention and Extinction of Fire (CTIF) as a joint project between France, Poland and Finland to address the issues of fire fighter safety with regard to such fire instances [15]. The emphasis of the project is upon how to design and construct buildings with an emphasis on providing the opportunity to safely conduct rescue and extinguishment operations.

4.1.3 Middle East and Asia

Fire researchers in the Middle East are also facing the new fire protection hazards introduced by modern warehouses. Growth in the shipping industry in the Middle East, particularly Dubai and Abu Dhabi, have placed greater emphasis upon the need for enhanced protection options. Craig Nixon at Tyco Fire and Security UAE has reported that the increased building heights for warehouse facilities have made previous detection technologies out of date. He has also stressed the importance of designing for fire protection during construction, rather than trying to retro-fit options afterwards [16].

The emphasis on warehouse fire protection in the Middle East has been upon passive fire protection, through the elimination of ignition sources and the containment of fires. They have greatly advocated the use of fire compartments and barriers, sprays and paints, fire stopping, fire-rated partitions and ceilings, and protection of steel construction materials. The drive has been to prevent the total collapse of warehouse buildings in order to allow for better fire fighting [16].

Additional emphasis has been placed upon the importance of training warehouse staff to identify fires and follow correct procedures, including notification and basic fire fighting techniques [16]. Current campaigns exist to educate businesses about safe operating practices and requirements and allocating additional resources toward conducting regular safety inspections in order to limit the occurrences of warehouse fires [17].

Several organizations have authority to implement civil defense regulations, including the Environment, Health and Safety (EHS) Department in Jafza. These organizations and the prescriptive design requirements vary by location. Currently, Abu Dhabi is adopting the International Fire Code (IFC) for protection of storage facilities [16].

In Japan, large warehouse fire protection has focused upon the use of automatic sprinklers including ESFR, high expansion foam suppression systems, utilization of positive pressure ventilation techniques, and the use of robotically controlled remote firefighting vehicles. Planning is also emphasized, as companies greater than 30 persons require fire protection managers, and all employees are trained in fire prevention, fighting, and rescue [18].

Stringent sprinkler installation requirements exist in Japan, such that warehouses larger than 700 m^2 (7,500 ft^2), or with rack storage taller than 10 m (33 ft) are required to have automatic sprinkler systems. All new warehouse construction is required to have ESFR sprinklers. In addition, several local laws enforce more stringent requirements than the national regulations [18].

Japanese warehouse properties are required to have pre-planned fire fighting strategies. These plans require a determination of the stored commodities, existing fire protection equipment, recommended firefighting tactics, as well as the required number of firefighters and firefighting equipment to provide the minimum proper response. In general, however, offensive fire fighting tactics are not considered effective for warehouse fires, and defensive tactics and building ventilation are often prescribed in such plans [18].

4.2 Fire Loss Incidents

Despite continued efforts to reduce the frequency and severity of warehouse fire incidents, numerous catastrophic events still occur throughout the world. Numerous organizations exist that attempt to compile and analyze the data from such fire events and develop statistical relationships between the causes and circumstances of various fires. In addition to statistical data, the specific circumstances of several large warehouse fire events are also available. The analyses of the fires occurring recently in large warehouse facilities are presented here.

4.2.1 Statistical Analysis

In the United States, statistical fire data is compiled by the National Fire Incident Reporting System (NFIRS). Such data is compiled from voluntary fire incident reports submitted by responding fire departments throughout the country. This data identifies the type of buildings involved in fires, the factors leading up to ignitions, the detection, suppression, fire fighting operations, as well as extinguishment procedures, and the losses of life and property resulting from such fires.

According to the statistical analysis, U.S. fire departments responded to an average of 1,350 warehouse structure fires annually during 2003–2006. An annual average of 21 civilian injuries, 5 civilian deaths, and $124 million in property damages have resulted from such fires [20]. These numbers do not include incidents occurring in cold storage properties, but according to EIA statistics, these properties only represent about 2 % of all warehouse storage facilities [4]. It is generally recognized that deaths and injuries are not a major part of the fire problem from warehouse fires, but rather that the costs of property damages are considerably greater than for other structure fires [20].

Size is a contributing factor in the risk of fires in warehouse properties. When warehouse fire incidents are grouped by building area, the greatest number of fire incidents are shown to occur in larger buildings, with buildings greater than 1,860 m^2 (20,000 ft^2) accounting for 26 % of all warehouse fires and buildings 93–465 m^2 (1,000–4,999 ft^2) accounting for 19 % of all fires [21]. According to the EIA building use statistics, nearly 45 % of all warehouse properties fall into the 93–465 m^2 (1,000–4,999 ft^2) area range, while only 15 % are greater than 1,860 m^2 (20,000 ft^2) [4]. The disproportionate number of fires relative to the building percentage (e.g., 26 % of fires for 15 % of buildings) is evidence of the increasing fire risks associated with the construction of larger warehouse facilities. The increase in fire incidents for larger warehouses may be attributable to an expected fire incident per square foot, and thus more square footage would imply greater occurrences of fires. Regardless of the cause of the disparity, protection for large area warehouses should consider the increased fire risks.

The common ignition locations for various warehouse fires were determined to be storage and shipping and loading areas. Unclassified storage areas (13 %), shipping receiving or loading areas (8 %), and storage rooms, tanks, or bins (6 %) were among the most frequent ignition locations [19]. The material first ignited in warehouse fires was most often sawn wood (12 %) and plastic (9 %). The material first ignited causing the most property damages; however, were cardboard (11 %) and fabrics (9 %) [21].

The ignition of warehouse fires was most commonly the result of electrical distribution (14 %) and intentional arson (13 %). However, the arson fires contributed most significantly toward the damage of property, and are the cause of 21 % of all warehouse fire damages [19]. The complete breakdown of the causes of warehouse structure fires determined by NFIRS and NFPA Surveys is shown in Fig. 1.

In general, the materials on-site during warehouse structure fires varied greatly and are fairly well distributed among various commodities. The materials involved

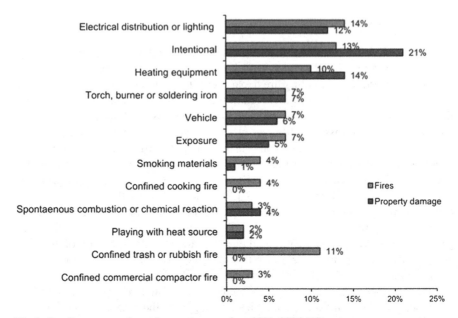

Fig. 1 Leading causes of warehouse structure fires 2003–2006 [19]

with the greatest number of fire incidents are paper products (10 %), wood (9 %), and motor vehicles (7 %). All other materials, including packing materials, flammable liquids, or electronics are generally involved in 2–5 % of fire incidents [21].

When considering the impact of materials upon the property damage costs, several materials contributed a large share. Containers or packing materials were involved in only 3 % of warehouse fires, but contributed 33 % to the property damages. Similarly, appliances and electronics accounted for 3 % of incidents, but 28 % of property damages [21].

Automatic detection equipment was present at only 22 % of warehouse fires, and was able to operate in only 15 % of all warehouse fires from 2005 to 2009 [21]. There was no detection equipment at 57 % of all warehouse fire incidents, while the remaining 21 % were confined fires that resulted in no measurable property damages. One unusual aspect of the role of detection equipment is that although equipment was present in 22 % of fires, those fires were attributed to 47 % of the property damages. This is possibly due to the increased use of detectors in warehouses containing valuable materials, but this cannot be determined from the statistical data.

When present, detection equipment is most often smoke detection devices, accounting for 36 % of all present detection equipment. Sprinkler water flow activation switches are also common, accounting for 27 % of present detection equipment. The presence of more than one type of detection is also identified in 18 % of scenarios, and it is likely that many of these events represent the combination of smoke detection and water flow switches [21].

From 2005 to 2009, automatic extinguishing systems were present at only 36 % of warehouse fires (excluding cold storage). This number has steadily increased from 13 % in 1980–1984 and 22 % in 1994–1998. In the cases where an automatic sprinkler was present, the system was a wet-pipe system in 82 % of cases and a dry pipe system in 17 % of cases. Other types of systems were only present in 1 % of occurrences [22].

There are additional factors influencing the costs of fires, related to suppression factors or circumstances that contributed to the growth and spread of the flames. Despite occurring in 10 % of warehouse incidents, a roof collapse is involved in 38 % of the fire damages. A wall collapse is a factor in 2 % of all incidents, yet is a factor in 25 % of all property damages. Another significant factor in fire cost is an inadequate water supply, occurring in 2 % of incidents but being a factor in 12 % of all fire damages [21]. These occurrences can frequently result when fires have grown too large and hazardous to risk entry for firefighting.

Similar warehouse fire reports have been compiled throughout the world. In the UK in 1999, there were 5,637 "Retail Distribution" premises fires and 4,678 "Industrial and Transport" premises fires. Although not directly reported, it is estimated by the United Kingdom Warehouse Association (UKWA) that 25 % of these fires were at warehouse facilities, or roughly 2,500 incidents per year [11]. Among these events, the Fire Protection Association (FPA) has found that 15–25 very large warehouse fires occurred annually from 1994 to 1999, resulting in £6–20 million in damages annually. Among these fire incidents, it was determined by the Large Fire Loss Index that arson is the most frequent cause of such fires, attributed to 46 % percent of warehouse fires. Electrical failure is also recognized as accounting for 17 %, while other factors such as hot work, spontaneous combustion, and smoking materials all contributed smaller portions [14].

Among all non-residential fires in the UK, including warehouse fires, no automatic fire detectors were present in 62 % of scenarios. Such devices were present and able to operate properly in 27 % of such fires, while failing to operate in 8 % of incidents [23].

The growing problem of warehouse fires has also been recognized in the Middle East. The Dubai police department has reported a 22 % increase in warehouse fires from 2009 to 2010. There were 31 warehouse fire incidents in 2009 compared to 38 in just the first 10 months of 2010. Seven percent of all fires occurring in Dubai are in warehouse properties. Investigators believe that the increase in fire incidents was caused by the increase in the addition of mezzanine levels, exceeded design storage capacities, the use of illegal partitions used to create additional work rooms, and from automatic extinguishing systems not being kept in working order [17].

4.2.2 Fire Incidents and Narratives

In addition to broad statistical data, further information about the role of building design, fire detection and suppression equipment, and fire department tactics and response can be determined from examination of specific warehouse fire incidents.

With thousands of incidents occurring globally every year, a select group of some recent incidents has been selected for discussion due to factors related to ignition, detection, or suppression.

One such fire occurred in a very large warehouse in New Orleans in 1996. The 380 × 278 m (1,245 × 912 ft: 1,135,440 ft^2) warehouse was filled with wicker baskets, rugs, polyfill pillows, cardboard boxes, towels, stacked plastic chairs, and various plastic bags and packaging. The facility contained 30 overhead and 17 in-rack sprinkler systems, but had a very high 22 m (72 ft) ceiling with a 16 m (53 ft) gap between the top of storage and the ceiling sprinklers. The storage system was operated using an ASRS system with a motorized cart [24].

A fire ignited in the stored materials and was able to set off the sprinkler system which provided notification through a water flow switch to the fire department. The large height of the building and the gap between the sprinklers and the stored materials rendered the sprinkler relatively ineffective at containment, but the fire department arrived in 8 min and took nearly 6 h to extinguish the fire. At that time, the valves controlling the sprinkler system were manually deactivated [24].

At that time, the ASRS was reactivated without inspection and a damaged wire arced and reignited the fire. At this time, the sprinkler system was still deactivated and the valves could not be restored in time to prevent significant fire spread. After the second ignition, the fire departments could not extinguish the fire for 6 days and the building was a total loss [24].

This fire demonstrates several important factors to consider in warehouse fires [24]. Despite a rapid response to the water flow alarm, the fully code compliant sprinkler system was not sufficient to properly contain the fire to allow fire department extinguishment operations. Secondly, the use of the ASRS presented an additional fire hazard when reactivated after the fire. Additionally, the sprinkler system had not been reactivated prior to activating the electrical equipment. The importance of restoring the function of the sprinkler system and inspecting electrical equipment before reactivation was key in this fire incident.

In the New Orleans warehouse fire, deactivating the sprinkler system after the fire was believed to be extinguished resulted in the total loss of the property. Manually deactivating the sprinkler systems during the fire to improve visibility is another tactic used by fire departments that has also resulted in total building losses. In general, smoke accumulates within a hot layer along the ceiling of a burning building, but smoke cooled by sprinklers is less buoyant and the sprinkler spray acts to pump smoke down to the floor, and often fills the space with smoke at floor level, reducing total visibility. Shutting down the sprinkler system to improve visibility is a tactic that has been employed to improve visibility [25]. Use of this method has resulted in significant fire losses, including contributing to the complete loss of a Tupperware warehouse in South Carolina in 2007.

This specific fire event consisted of a 15,329 m^2 (165,000 ft^2) warehouse containing plastic Tupperware in cases in rack storage in four multiple row racks with 16 sections each 11.6 m (38 ft) high, including 172 lanes stored 29 totes deep. The building was fully protected with a code compliant sprinkler system, including in-rack sprinklers with horizontal barriers at tiers 4, 8, and 12 that activated properly during the fire, but were unable to extinguish the fire [26].

Upon arrival, the reduced visibility from smoke prevented the fire department from locating the seat of the fire and properly attacking the fire. To improve visibility, the sprinkler system was shut down for 20 min and a side wall was removed to improve ventilation. A fire located deep in the building and high in the racks prevented the fire department from extinguishing the fire, and the increased ventilation and lack of sprinklers drove the fire out of control and the building was a total loss [26]. As evidenced by this event, even fully sprinklered and protected buildings are subject to total fire loss. The tactic of shutting down the sprinkler system to improve visibility is now explicitly prohibited by NFPA 13E, Sect. 4.3.3 [27].

Several other fire incidents have been well documented by NFIRS narratives constructed by the responding fire departments. One incident occurred in California in 2008 at a warehouse storing almonds in wooden bins when a water reactive fumigant was exposed to moisture and ignited. There was no detection or suppression equipment present and the fire was able to spread within the warehouse until it was observed by a sheriff who notified dispatch. By the time fire departments were able to respond the building was a total loss, resulting in over $7.5 million in damages.

Failure to detect and provide notification has been a significant factor in several other large warehouse fire incidents. A 2009 Montana fire ignited in a tire warehouse of wooden construction. No detection or suppression equipment was present and the fire was not reported until a passerby called 911. At the time of arrival, the fire was too large to engage in anything other than defensive fire fighting actions and the fire resulted in a roof collapse and a total building loss of over $2 million [28].

A similar loss occurred at a 2010 Wisconsin warehouse containing stacked cardboard. There was no detection or suppression equipment and the fire was not reported until smoke was smelled by a neighbor. The fire department was unable to locate the seat of the fire and the fire took over 18 h to extinguish causing $3.5 million in damages [29].

Even when suppression systems are present, delayed fire department notification can result in catastrophic fire events. A 2011 Illinois fire in a warehouse containing palletized plastic bottles in plastic shrink-wrap was started by arson. The sprinkler system activated, but the water flow switch on the system was not monitored, and thus no one was notified of the event until workers in an adjacent building alerted the fire department. The fire department was only able to ventilate the roof and attempt to support the sprinkler system, but the fire resulted in $1.5 million in damages [30].

A similar occurrence affected a 2009 Pennsylvania fire when a vehicle fire spread to adjacent warehouse spaces. Although no suppression system was present, the building did contain heat detection. However, the monitoring station failed to report the alarm to the fire department, and the fire was not reported until a nearby police car observed the blaze. The fire took 3 h to control and resulted in a total building loss of $4 million.

The proper detection, notification, and suppression system responses have protected warehouse buildings from becoming large fire events. Conversely, adequate response, planning, and fire protection design have still resulted in events with significant fire losses. In general, many large fire events are the result of a delay in the notification system, the improper design of fire protection systems for the building hazards, or the inactivation or improper operation of suppression systems.

An analysis conducted of fire incidents conducted in 2006 determined several key lessons [31]. In general, analysis of fire incidents confirms that sprinkler design must match the hazards. It was observed that automatic detection and notification was a key element in protection, even when performed by a water flow switch. When automatic detection was not present, the importance for occupants and personal to be well trained and aware of the proper response to fire events was also stressed. It was observed that manual fire fighting attempts often failed and cost the fire department valuable response time. Additional observations confirmed that excessive water application for sustained firefighting efforts could result in toxic runoff.

Recommendations for prevention of warehouse fire events were directed toward security to prevent arson, the application of well documented hot work policies, effective planning to prevent incompatible storage, and vigilant equipment maintenance programs. The benefits of compartmentation in preventing fire growth and spread were also cited as key components [31].

4.3 Applicable Detection Systems

Various technologies exist that have been used or could potentially provide early warning fire detection for warehouse environments. Several technologies are discussed in this section, including the basic operating principles and general warehouse environment applicability. The discussion of benefits and limitations involved for various technologies within this section is kept entirely conceptual, and some realistic detection results of warehouse related applications are discussed in Sect. 5.

In general, a fire can be detected by the presence of smoke, heat, or flames. Sprinkler actuation currently serves as a form of heat detection, as the sprinkler is activated by excessive heat and the sudden flow of water can be monitored and used to provide notification of a fire event to occupants and/or fire departments. In the absence of sprinklers, fires are often detected by the visible presence of smoke or flames to occupants or a passerby. Due to the high potential for rapid fire growth, these methods often are not capable of detecting fires early enough to provide a fire department adequate response time to prevent large fire loss events. Other available technologies may be capable of faster detection times and may reduce the overall fire fighter response times. The suitability of certain types of detection will depend upon the size of fires they are intended to detect, the duration of the incipient period of the fires, the fire growth rate after incipient development, and the size and configuration of the space they are designed to protect.

4.3.1 Smoke Detection

Smoke detection is often considered a reliable option for early warning fire detection. Smoke is produced in the early stages of fire development, often long before the initiation of rapid flame spread. Smoke is released from the source of ignition and can travel through heat-induced buoyancy or forced air flow. There are several technologies capable of smoke detection in warehouse environments.

Traditional smoke detection systems involve the use of spot type smoke detectors. Spot detectors refer to devices that can be mounted to ceilings or walls and measure the concentration of smoke at a single point. Typical commercial spot detectors used are of the photoelectric type.

While spot type smoke detectors are often capable of detection of fires more rapidly than sprinkler activation, they are still dependent upon the migration of smoke particles to the location of the detectors. Generally, the migration of smoke occurs due to the buoyancy of the heated smoke. In large warehouse facilities, especially where the ceiling or detector location is excessively high, the buoyancy of smoke from small incipient fires may not be sufficient to drive the smoke to the detector locations. Besides deficiency in detection capability in these facilities, spot smoke detectors are not ideal due to maintenance and cost issues.

Periodic cleaning may be a necessity to maintain detector sensitivity and avoid trouble and false alarms. Smoke detectors require a semiannual visual inspection and an annual check for smoke entry and alarm in accordance with NFPA 72 [32]. Many addressable spot smoke detectors are capable of performing electronic self tests for cleanliness and sensitivity, and thus only the annual smoke entry test must be physically performed. Therefore maintenance, inspection and testing require-ments can be overly burdensome and costly given the logistics of accessing devices on high ceilings with stored commodities.

Optical beam smoke detectors provide a means to span larger distances across open spaces by providing detection along a light beam between a source and receiver. These devices utilize the principle of obscuration of the light beam to detect smoke that crosses the path of the beam. The coverage of a beam detector ranges from upward of about 50 m to 150 m (160–500 ft). Beam detectors are often used for the protection of large open spaces due to the ability to detect over long distances and protect large areas with a limited number of detectors. Similar to spot detectors, beam detectors still require smoke to migrate to the area of detection (i.e., through the beam). Where high ceilings are of concern, installation at such heights may become an issue. The devices can be placed at intermediate heights across the warehouse, but would need to compensate for the movement of equipment, such as ASRS, through the detection beams. Other issues that can affect beam detectors are misalignment due to building motion and vibration. Many warehouse facilities are constructed of steel framing and columns and such construction is susceptible to motion due to wind or even moving equipment. Different beam technologies have a range of susceptibility and compensation/warning to device motion and misalignment.

Smoke detection can also be performed actively through air sampling. Air aspiration smoke detectors (ASD) draw air from holes in a tube or pipe run through the protected space. This sampled air is then run back to a central detector and analyzed for smoke.

Air sampling systems (a.k.a. air aspiration) typically have a wide range of sensitivity, encompassing typical ranges of spot smoke detectors and including a much more sensitive range with smoke alarm levels over an order of magnitude lower. While these systems draw air to the detector, the extremely low flow only pulls air from a short distance from the sampling port; thus air sampling systems are still dependent on buoyancy or mechanical ventilation to move smoke to the sampling hole locations. The sampling holes can be widely distributed to provide a vast area of coverage, both at ceiling level and intermediate heights. Aspiration systems are often used in high airflow environments. They can be used in dirty environments by maintaining a measured level of ambient particulate as a reference and with the use of proper filters.

Video image detection (VID) systems consist of video-based analytical algorithms that integrate cameras into advanced flame and smoke detection systems. The video image from an analog or digital camera is processed by proprietary software to determine if smoke or flame from a fire is identified in the video. The detection algorithms use different techniques to identify the flame and smoke characteristics and can be based on spectral, spatial or temporal properties; these include assessing changes in brightness, contrast, edge content, motion, dynamic frequencies, and pattern and color matching [33].

VID devices can be operated as a system of distributed cameras monitored by an external computer station or individually as single camera detectors. The smoke alarm signals can be monitored independently for the fire alarm system, while the video signals can be simultaneously used for security, false alarm prevention, and determination of fire fighting tactics [33]. With arson identified as a common cause of warehouse fires, additional security and monitoring is of key concern.

While VID smoke detectors do require smoke to enter the field of view, they are able to cover a larger volume of a space compared to spot or line-type (i.e., beam) detectors. This provides significant benefits for large protected areas with high ceilings over spot, beam, and aspiration type smoke detection. Smoke detection VID systems require light to view the rising smoke plumes, but can use IR illuminators or IR sensitive cameras to increase sensitivity [33]. Despite broader coverage, the structures within a warehouse can create difficult coverage relative to sensitivity and potential nuisance sources.

The overall sensitivity of VID smoke detectors is a function of the visual image, and could be affected by background images, colors of walls and objects, lighting, and activities in the warehouse. The understanding and prevention of false alarms is a key element of VID performance due to variation in detector algorithms. Every installation location may have unique features that could affect the system performance, depending on manufacturer. Some potential sources for nuisance alarms include moving objects, changes in ambient light, heat sources, hot work, steam, dust, or lens contamination or condensation. On-site validation of the VID devices may be necessary due to variability in outside factors [33].

4.3.2 Heat Detection

In addition to smoke detection, warehouse fires can be detected by measuring the temperature of hot gases or surfaces exposed to fires. There are generally two types of heat detection methods utilized, spot detection and linear detection. Traditional heat activated sprinkler protection systems with water flow monitors can be considered as a form of spot heat detection. Fires can be detected through heat detection by measuring fixed temperature or rate-of-rise temperature thresholds.

Spot heat detectors can be addressable or conventional, and can be used to provide audible notification to occupants and electronic notification to monitoring stations and fire departments. Like sprinkler heads, spot detectors can be placed at the ceiling or within storage racks to measure elevated gas and surface temperatures resulting from fires. It is possible with some addressable detectors to also monitor temperature, which could be used as a pre-alarm condition for lower level temperature increases.

Linear heat detection (LHD) describes a general type of cable or tube based detector that can measure heat (temperature) along the length of the device. There are several principles of operation for linear heat measurement. The primary types include continuous thermocouples, heat sensitive cables, electrically conductive, fiber optic, and pneumatic. The physical operating principles of each type vary, but in general they are extended wires or tubes that are capable of sensing elevated temperatures anywhere along an extended path.

Linear heat detection is often advantageous for warehouse facilities because it can be run in storage racks and between storage levels. Linear heat detection devices can be very rugged, resistant to environmental conditions and low maintenance. Placement on or near storage provides linear heat detection with the benefit of much faster alarm times compared to ceiling mounted detection devices. However, heat detection will only detect developing flaming fires and may not provide notification during incipient periods prior to rapid flame growth. The impact of the fire development with regard to detection is evaluated in Sect. 5.

4.3.3 Flame Detection

Heat detection devices measure the convective output from hot gases in a fire, the presence of flames can also be measured by the radiative components. Flame detection can be performed through optical visualization of protected spaces by measuring emitted thermal radiation or through resolved imaging technologies.

Optical flame detectors (OFD) measure radiant energy emitted by flames at specific wavelengths indicative of fires. For example, a burning hydrocarbon fuel produces hot CO_2 which has a strong emission of radiation at 4.4 μm. Various OFDs are designed to detect ultraviolet (UV), infrared (IR), or combinations of various wavelengths of the radiative spectrum. The wavelengths detected can be tailored to detect specific types of flames and reject other forms of background or nuisance sources.

A common form of OFD currently utilizes a multi-spectrum IR approach, whereby the device detects multiple unique IR wave bands simultaneously. The ratios of the energy in the various measurement bands are used to make determinations about flaming fire events. Extremely strong radiation detected at the CO_2 emission wavelength compared to other reference wavelengths may indicate the presence of a fire, for example. An additional criterion for fire events is measured in the transient flicker of the incoming radiation. It is known that flames flicker at a frequency of 1–10 Hz. Detection of this type of oscillation may indicate the presence of a fire. The multi-spectrum IR detection algorithms uniquely combine the spectral and temporal profiles of incoming radiation and make a determination about the presence of a fire.

OFD devices contain lenses and can be used to monitor a large area for the presence of flames, many devices are capable of protecting a conical area ±45 degrees in the vertical and horizontal extending a distance of 30 m (100 ft) or more from the actual device. Such devices could be used to protect a large area, such as a warehouse facility. The primary limitation of OFDs, is due to the presence of obstructions. The devices must be able to "see" the flames, and warehouses generally present a fairly dense viewing environment. A large number of OFD devices may be necessary to visualize the extent of the spaces between stored materials in a warehouse.

Another type of visual flame detection technology is VID as noted above. In addition to being used for detection of smoke emissions, these devices can be programmed with algorithms for the detection of flames. While VID flame detection is limited by the presence of obstructions like OFD, they do provide the added advantage of full video image visualization. Rather than integrating the incoming signal, the VID detectors can provide operators with a fully resolved image of the protected area, allowing identification of the seat of the fire, involved commodities, or potential false alarms. Identification of the exact location of an incipient fire can provide valuable time to fire fighting operations in the prevention of large fire events.

4.4 Warehouse Detection Research

Several projects have been undertaken to determine the applicability of various fire detection technologies to warehouse or similar applications. This includes facilities with very large floor areas or ceiling heights, or facilities with large amounts of visually obstructing materials. The test conditions, results, and general conclusions of several such tests are presented and reviewed here.

A full scale detection test was conducted in Australia in 2008 to determine the applicability of spot, aspiration, and beam smoke detection systems for large warehouse type facilities with high roofs and large floor areas. The challenges by such spaces for smoke detection were identified as the high ceiling heights, leaky and drafty construction, and the large total volume of air in the space.

In addition, forced air ventilation was identified as a potential source of dilution that would restrict the detection of smoke in the space [34].

The test space consisted of a large warehouse 43 × 12 m (141 × 40 ft) with a ventilated ridge ceiling spanning from 8–8.5 m (26–28 ft) at the apex. Spot smoke detectors were placed next to individual aspirating smoke detector sampling holes at a height of 7.2 m (24 ft) and the beam detector was placed just above at a height of 8 m (26 ft).

Fire sources tested included a 1–2 kW wooden timber fire, a 10–20 kW heptane pool fire, and a 5–6 kW smoldering pellet ignited by a radiant heater. The aspiration detector alarmed to all timber and heptane pool fires, and to 4/5 of the smoldering pellet tests. The spot smoke detectors were able to alarm to 2/5 of the smoldering pellet tests, and the beam detector did not produce an alarm to any fire test [34].

These tests demonstrated that for roughly equivalent installation siting conditions, the aspiration smoke detection system used for testing was able to alarm to more fire sources for the high ceiling test facility. However, the individual alarm thresholds and sensitivities for each device are not reported. In addition, the tests did not account for the potential rapid fire growth resulting from typical warehouse and storage applications, but rather measured the detection performance of small, incipient fires with little to no potential for growth.

The absence of stacked commodities also influenced the overall smoke behavior. While the test was intended to identify the performance of the smoke detection systems for warehouse environments, it is possible that the presence of such commodities could influence the buoyancy and migration patterns of the developed smoke, further influencing detection performance. In general, these tests provide a fair representation of smoke detection systems ability to detect small fire sources located on the ground of a facility with a moderately high ceiling and large volume.

Another set of full scale tests were developed to investigate the performance of aspiration smoke and optical beam detection in facilities with high ceilings in Europe in 2010 by the BRE. These tests also did not utilize large amounts of stored materials or demonstrate fires with rapid growth rates or potentials. They similarly investigated the ability of the detection systems to alarm to small fires located far below detectors but with larger ceiling heights and additional detector installations and sensitivities [35].

The test facility consisted of a 64 × 40 m (210 × 131 ft) enclosure with a sloped roof height from 41–43.5 m (135–143 ft), sloped along the 40 m (131 ft) width of the building. Test fires were located halfway along the 64 m (210 ft) length of the building and one quarter along the 40 m (131 ft) width, such that the fire was located below the middle of the ceiling slope, halfway to the apex. The vertical temperature gradient in the space was approximately.

0.18–0.28 °C/m, or a total of 12 °C from top to bottom with no cross air flow or ventilation. Temperature measurements were made at various heights above the fire and along the ceiling using thermocouples [35]. In all test scenarios, the temperature rise at the ceiling was almost negligible, indicating the limited applicability of ceiling heat detectors (or sprinklers) for detection of the tested fire sources (described below).

Aspiration smoke detectors, conforming with EN54-20:2006, were installed for Class A, B, and C sensitivity levels. The three alarm thresholds are not identified by the manufacturer, but each Class falls within the prescribed allowable standards. Beam detectors, also conforming to EN54, were installed with 35 % obscuration alarm settings [35].

One beam detector was installed spanning the 40 m (131 ft) room width at a height of 16.5 m (54 ft). Another was located directly above the fire angled along the slope of the ceiling. Two other beam detectors were placed spanning 36 m (118 ft) in the center of the room running lengthwise, with one placed at the apex of the ceiling and the other located 10 m (33 ft) down the slope of the ceiling. Such installation was consistent with British Standard 5839-1, allowing separations of 15 m (49 ft), with no more than 7.5 m (25 ft) to a wall and 0.6 m (2 ft) to the apex.

The aspiration system was installed in three configurations. A pipe was installed with a single hole placed directly above the fire location at the ceiling; this was installed to represent a spot type smoke detector. A single pipe system with 5 holes each spaced 4 m (13 ft) apart was installed running along the slope of the ceiling at the center of the room also directly above the fire. A complete ASD system was also installed consisting of 4 pipes with 5 holes on each pipe spaced 8 m (26 ft) apart, oriented along the slope of the ceiling with the fire placed in the middle of two pipes. The system was designed in accordance with British Standard 3839-1, such that no point below the pipe array was located more than 7.5 m (25 ft) from a sampling hole [35].

Numerous fire tests were conducted to analyze the performance of the smoke detection systems. A wood crib fire, approximately 220 kW in size was conducted. This test provided heat but a minimal amount of smoke. These tests did not produce an alarm in any of the beam detectors, reaching only half the alarm threshold, and produced one alarm in the highest sensitivity aspiration detector [35].

A larger wood crib fire was also conducted, producing a fire approximately 560 kW in size. This fire produced an alarm in the beam detector at 16.5 m (54 ft) height, but no others. The 4-pipe aspiration smoke detector alarmed at both Class A and B sensitivities, and the 1-pipe and 1-hole aspiration detectors alarmed at Class A sensitivities [35].

A heptane pan fire, approximately 400 kW, was conducted to produce a large amount of heat and smoke. The fire produced alarms in the ceiling and apex beam detectors, but not the intermediate height location. The 4-pipe ASD alarmed for all three sensitivities, and the 1-pipe system alarmed at the highest sensitivity [35].

A fire consisting of crumpled newspapers in a steel bin of approximately 30 kW was conducted to produce a large amount of white smoke, followed by a high heat flaming fire. The beam detectors alarmed at the ceiling, apex, and intermediate heights and the ASD detectors alarmed for all orientations and sensitivities except the single hole, lowest sensitivity device [35].

A mixture of potassium chlorate and lactose were also tested to produce a highly buoyant white smoke. This fire was detected by all the beam detectors with the exception of the apex detector. The ASD detectors alarmed for all orientations and sensitivities except the single hole, lowest sensitivity device [35].

Lastly, a fire consisting of smoke pellets over a butane burner was examined, also producing a moderately buoyant white smoke. The intermediate height beam detectors alarmed, but none at the ceiling level due to the limited buoyancy of the smoke. Only the 4-pipe Class A ASD system was able to detect this fire at the ceiling level [35].

In general, both the beam and aspiration detection were able to detect a wide variety of fires and smoke types with higher sensitivity levels. Decreasing the beam sensitivity to 50 % obscuration would have resulted in no alarms, and increasing to 25 % would have produced alarms in nearly all test cases. Both levels are allowable by standards [35].

The beam detectors placed at intermediate heights were most effective at detecting fires producing large amounts of smoke with limited buoyancy, and the aspiration detector with the 4-pipe sampling, a large area, was in general the most effective detection method [35].

The test investigators did recommend additional investigations into potential factors influencing the performance of aspiration and beam detection in warehouse applications. They recommend investigating:

• Background levels to determine sensitivity to avoid false alarms
• More challenging fire locations
• Fire sources producing black smoke
• The influence of cross flow air in space
• Thermal gradients existing in high ceiling facilities
• Performance of dropped pipes (not at ceiling from ASD).

In addition to the performance of spot, aspiration, and beam smoke detection, recent testing has also been conducted to investigate the performance of VID smoke and flame detection for large area and height facilities. One test conducted was oriented toward the performance of such devices in atria rather than warehouses, but did investigate the performance of such devices in the presence of obstructions, a key element of warehouse detection performance.

An 18.9 × 18.4 m (62 × 60 ft) building, 26.4 m (86 ft) in height, was used to test the performance of smoke and flame detection using VID technologies. VID devices were placed in various locations, including at low and high elevations and with multiple FOV settings. Testing consisted of alcohol, heptane, and JP8 pan fires, an n-heptane spray fire, paper and wood fires, burning denim jeans, and a smoke bomb. Fires were also viewed partially obstructed by a table, a plate, and clothing racks. In addition, lens contamination was investigated by placing a metal screen and cloth in front of the cameras to reduce incoming light signals. The devices were also tested for false alarm prevention through exposure to chopped lights and heat sources [36].

The VID devices were able to detect various obstructed and unobstructed flame and smoke scenarios quickly during the early stages of fire growth. In general, the unobstructed fires were unaffected by changes in the fire size or fuel loading, but larger fires did result in faster detection for obstructed scenarios. Tests conducted with a cross flow of air also produced faster detection times than those with still air, perhaps due to enhanced smoke migration. The VID detectors were also resistant to the tested false alarm sources, including IR and bright light sources [36].

The placement of the VID devices appeared to have the greatest impact upon system performance. In general, the VID devices located higher achieved better detection rates and frequencies than when placed lower in the facility. It is possible that this was due to the increased reflections of the fire sources from the metal walls of the test structure opposed to the concrete floor as viewed from above [36]. It may also be that the higher devices had better views of the fires directly.

The use of the obscuration wire mesh and cloth did not significantly impact the detection of flames, but did produce slightly blurry images and reduced the ability of the devices to detect smoke [36].

Testing of various detection technologies for warehouse environments was performed in 2011 by axonX, a manufacturer of smoke and flame detection VID systems, and Schirmer Engineering. This test series was designed to evaluate the relative performance of VID, ASD, beam detection, spot ionization and photoelectric smoke detectors, as well as linear heat detection. The test facility consisted of a laboratory space 17.6 × 10.6 m (59.6 × 34.8 ft) with a ceiling height of 5.5 m (18.1 ft). All of the doors were closed to prevent air flow through the space and the windows and skylights were blacked out to prevent variable lighting conditions [37].

An ASD system was installed consisting of a single pipe with three evenly spaced sample points running along the 17.6 m (59.6 ft) length of the room. The ASD system was configured to monitor two separate alarm obscuration levels, including a Fire 1 setting at 0.2 %/m (0.06 %/ft) and a Fire 2 alarm at 2.0 %/m (0.6 %/ft).

A beam detector was installed at a height of 4.6 m (15 ft) with the beam directed along the 17.6 m (59.6 ft) dimension along the center axis. The beam emitter and receiver were each placed along the center axis a distance of 1.5 m (5.2 ft) from the walls, spanning a total beam length of 15.3 m (49.2 ft). The beam detector was set to an obscuration alarm level of 30 %, equating to a nominal alarm threshold of 1.94 %/m (0.6 %/ft).

Photoelectric and ionization spot smoke detectors were placed in two ceiling locations along the 17.6 m (59.6 ft) centerline of the space at a distance of 4.6 m (15 ft) from the end walls. The ionization detectors were set to an alarm sensitivity of 1.96 %/m (0.61 %/ft) and the photoelectric detectors were set to 2.0 %/m (0.6 %/ft). Linear heat detection was run at the ceiling in a single loop at a distance of 2.1 m and 2.4 m (7 ft and 8 ft) from the 17.6 m (59.6 ft) walls with an alarm temperature setting of 68 °C (155 °F). VID devices, consisting of both flame and smoke detection algorithms, were placed at heights of 4.2 m (13.7 ft) and 3.4 m (11 ft) in two adjacent corners of the space along the 17.6 m (59.6 ft) wall with a system alarm sensitivity of "Medium."

Tests were conducted using various smoke and flame sources. Tested smoke sources included black and white smoke emitters, smoldering wires, and smoldering wood. Tested flaming sources included a 50/50 mixture of toluene and heptane, a wood crib, and tamped paper in a metal container. The tamped paper produced a smoky, smoldering source followed shortly after by a flaming ignition. Test fire sources were evaluated in three different locations, including the center of the test space directly below ASD and spot detector locations, again below the detection locations but obscured from view by racks of boxes placed 2.1 m (7 ft) high, and in the corner of the test facility below a VID detector and away from the spot, beam, and ASD location [37].

In general, the VID detection system provided the fastest and most reliable alarm responses among the evaluated detectors. The flammable liquid and smoldering wood fires obscured by the racks of boxes were the only notable exceptions to this observation. VID alarms were obtained for these scenarios; however, the ASD with the Fire 1 alarm threshold and the VID detectors were the only technologies capable of alarming to all tested scenarios. All other technologies did not detect some of the tested fire sources. None of the evaluated test fires produced enough heat to allow detection by the linear heat detector [37].

All of the technologies were not capable of detecting the smoking fire sources. The ionization detectors were unable to detect the black smoke emitter when placed in the corner, and the photoelectric detectors alarmed nearly 6 min later than any other device during these tests. The white smoke emitter could not be detected by the ionization detectors for any test location, by the photoelectric detectors for the obstructed location, nor by the beam detector for the corner test location. The VID system responded most rapidly to the white smoke emitter, generally alarming within 3 min while all other devices produced alarms in 5–12 min. Only the VID and ASD Fire 1 were able to detect the smoldering wire in all test cases, and the VID detectors alarmed in 4–8 min while all other device alarms occurred between 12 and 30 min. The smoldering wood fires were not detected by the photoelectric detectors when placed in the open or the corner, nor by the ionization detectors when placed in the corner. All other devices produced comparable alarm times during the smoldering wood tests [37].

Similar results were obtained for flaming fire sources. The VID detectors produced the most rapid alarm response to the flaming wood crib in the corner, providing an alarm in approximately 1 min while all other alarms took 4–6 min. The beam and photoelectric smoke detectors were not able to produce any alarms to the wood crib in the corner. The VID also responded fastest to the tamped paper fires, producing alarms in 0.5–1.5 min while all other alarms took 2.5–5.5 min. The ionization and photoelectric smoke detectors were not able to produce an alarm to the tamped paper when located in the corner. All detectors were able to alarm to the flaming liquid pool fire. It should be noted that the obstructed VID detectors, the photoelectric, and the beam detectors generally took the longest of the technologies among pool fire tests when the fires were placed in the corner [37].

In addition to testing specifically for the evaluation of detection systems, several other investigations have been performed regarding warehouse fires. One such investigation focused upon the application of high expansion foam systems for extinguishing fires in modular stacked plastic bins with an ASRS system. The experiments did include the use of aspiration and beam smoke detection above the modular storage units and determined that in general, the aspiration smoke detection method was routinely faster than the beam detectors [38].

In addition to full scale experimentation, computer modeling has been used to investigate the issues of warehouse fire development and response. One study investigated the potential fire growth rates, available time to evacuate, and potential fire department tactics at the time of arrival. During the investigation, the model assumed that beam detection would be able to provide notification to occupants and the fire department at the instant of exponential fire growth in a cardboard and polystyrene packaging material. It is assumed that the beam detection is not capable of detection during the incubation period, but would detect immediately upon the development of a spreading fire [38].

The model used a warehouse fire model developed by the BRE that estimates that such a fire would be able to grow to the top of a 12.5 m (41 ft) tall stack within 2 min, producing a 15 MW fire size. Within 9 min, the horizontal fire spread was estimated to produce a fire of 130 MW. It was calculated that a 3–4 min evacuation time from the point of detection would be adequate to allow occupants to safely evacuate the space prior to reaching untenable conditions, but an estimated 8–10 min fire department response time would still not provide sufficient time to enter the building to actively fight the fire and prevent significant loss scenarios [39]. This fire scenario analysis indicates that even with fairly rapid fire detection, it may not be possible for fire departments to reach warehouse fires prior to significant flame spread and the development of conditions too dangerous to allow entry.

4.5 Literature Review Conclusions

The warehouse storage industry has been consistently working toward improvements in efficiency of design. These improvements have resulted in larger and taller warehouses, the use of automated storage and retrieval systems (ASRS), and increased storage densities and heights. While general efficiency has improved, additional fire protection threats have also been introduced. Increased storage heights and densities have increased the potential for flame spread and growth while limiting the potential effectiveness of detection and suppression systems. Such fire incidents limit the ability of fire departments to enter such facilities during fire events, contributing toward total loss fire incidents.

While the changes in warehouse characteristics are changing globally, not all regions and countries have approached the problem with the same strategies. Building regulations and statutory requirements can vary greatly by nation and even through local jurisdictions. Such regulations impact the design of buildings, the required usage of detection and suppression equipment, the required levels of employee education and fire event pre-planning, and the appropriate fire department response levels.

Fire incident data indicate that large warehouses are often at greater risk for fire occurrences, and the potential losses from such facilities are the greatest. Common causes of warehouse fires were arson and the failure of electrical equipment, often starting in storage and shipping areas. Wood, paper, plastic, and fabrics were the materials most commonly ignited and contributed to the large fire loss events. Varying globally, detection equipment was present in 20–40 % of all warehouse fire incidents and suppression systems were in 30–40 % of warehouse fire incidents.

Evaluation of large fire incidents indicates that early warning notification of fire incidents is often not-present or fails to provide the proper response to fire departments. In this case, early warning notification refers to an alert or indication of the fire condition prior to potential sprinkler activation. It is possible that given additional response time, improved firefighting tactics could reduce the number of such large fire events. Scenarios where detection may improve the response time and fire fighting effectiveness are investigated further in the next section.

Early fire detection options include the use of smoke, heat, and flame detection. Smoke detectors for use in warehouse applications can include photoelectric spot type, beam detectors, aspiration smoke detection, and video image smoke detection. Smoke detection often provides the earliest warning of fires, due to the potential presence of smoke prior to flaming ignitions and the ability of smoke to migrate through buoyancy and air flow away from the seat of the fire.

Research into the effectiveness of various detection technologies for multiple fire types have been performed for large spaces with high ceilings, similar to the challenges encountered in large warehouse facilities. It has generally been found that smoke detection can be effective for such facilities, with aspiration detection generally proving more effective performance than either beam or spot type detectors. Video image detection has also been shown to provide fast response to warehouse fire types, but issues of obstructions in warehouses may limit the effectiveness of flame detection VID systems. Computer modeling of warehouse fire incidents has potentially indicated that rapid fire growth and large combustible loading in warehouse facilities may prove too challenging for fire departments even given extremely early notification and fast response times. This point is evaluated further in the next section.

5 Fire Hazard Analysis

A fire hazard analysis has been developed to determine the potential for utilizing early warning detection equipment to improve fire department response times to fight warehouse fires and to reduce the overall costs of fire losses.

A time line describing the conditions of the fire and the stages of manual response is shown in Fig. 2. An effective detection system can perform two functions to improve the manual response and reduce the potential for property destruction. First, the detection system can reduce time of discovery and initiate the fire department response. This enhanced notification would reduce the severity of the fire upon arrival of the fire department.

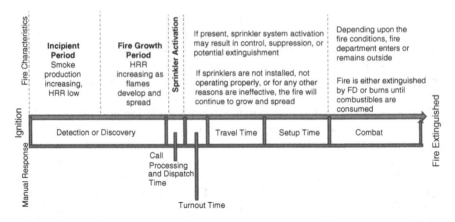

Fig. 2 Fire development and manual response timelines

Secondly, the detection system can provide additional information to the fire department upon arrival concerning the ignition location and may provide information regarding the extent and severity of the fire spread. While clearly valuable for size-up and setup upon arrival at the scene, the ability of a detection system to provide this information would be a characteristic of the specific detection system, and thus is not evaluated within this analysis.

5.1 Characteristic Warehouse Designs

Warehouse designs vary widely based upon the function and demands. The key variables affecting the fire performance of warehouses include the stored commodities and arrangements, the building height and area, fire protection features such as detection, suppression, or ventilation systems, and the presence of any additional equipment such as automatic storage and retrieval systems (ASRS). For this analysis, several warehouse designs are considered with a number of characteristics in common while other characteristics are varied for analysis.

The common base design consists of a 9,290 m² (100,000 ft²) floor area with unprotected steel construction. This area was chosen because it requires long travel distances for firefighting access while providing fast smoke layer development resulting in reduced visibility. Each warehouse consists of fixed double row rack storage. This is a common storage configuration with the potential for large and rapid fire growth. The storage of the racks will consist of longitudinal and transverse flue spaces between racks of 15 cm (6 in) in accordance with NFPA 13, Sect. 16.1.11.1.1 and 16.1.10.2 and minimum aisle widths of 1.2 m (4 ft) in accordance with NFPA 13 [3].

The primary variables investigated for the warehouse designs consist of the building and storage height and the type of the stored commodities. Six warehouse scenarios are considered for building heights of 4.6 m (15 ft), 12.2 m (40 ft), and 30.5 m (100ft) with a Class II and a Group A Plastic commodity (see Sect. 5.2.1 for definitions). The fire protection systems for each warehouse height and commodity are in accordance with NFPA 13 [3]. The warehouse designs are described below and summarized in Table 1.

5.1.1 Warehouse Design 1

Warehouse design 1 (W1) consists of a 4.2 m (15 ft) tall warehouse with 2 tiers of Class II storage, reaching up to a height of 3.7 m (12 ft). An automatic sprinkler system is installed at the ceiling in accordance with NFPA 13 Table 13.2.1 and Fig. 13.2.1. The sprinklers are spaced across the ceiling at 3.0 × 3.0 m (10 × 10 ft) spacing with an ordinary temperature response of 74 °C (165 °F) and an RTI of 80 $m^{1/2} s^{1/2}$ (145 $ft^{1/2} s^{1/2}$).

5.1.2 Warehouse Design 2

Warehouse design 2 (W2) consists of a 4.2 m (15 ft) tall warehouse with 2 tiers of Group A Plastic storage, reaching up to a height of 3.7 m (12 ft). An automatic sprinkler system is installed at the ceiling in accordance with NFPA 13 Fig. 17.2.1.2.1 (b) for storage of plastic commodities. The sprinklers are spaced across the ceiling at 3.0 × 3.0 m (10 × 10 ft) spacing with an ordinary temperature response of 74 °C (165 °F) and an RTI of 80 $m^{1/2} s^{1/2}$ (145 $ft^{1/2} s^{1/2}$).

5.1.3 Warehouse Design 3

Warehouse design 3 (W3) consists of a 12.2 m (40 ft) tall warehouse with 7 tiers of Class II storage, reaching up to a height of 10.7 m (35 ft) with an ASRS. An automatic sprinkler system has been installed at the ceiling in accordance with NFPA 13 Table 16.3.3.1 for storage of Class II commodities at the storage and building height specified. The sprinkler system consists of Early Suppression Fast Response (ESFR) sprinkler heads with quick response RTI of 50 $m^{1/2} s^{1/2}$ (91 $ft^{1/2} s^{1/2}$) at 74 °C (165 °F). No in-rack sprinklers are provided due to the use of ESFR sprinklers.

5.1.4 Warehouse Design 4

Warehouse design 3 (W3) consists of a 12.2 m (40 ft) tall warehouse with 7 tiers of Group A Plastic storage, reaching up to a height of 10.7 m (35 ft) with an ASRS. An automatic sprinkler system has been installed at the ceiling in accordance with NFPA 13 Table 17.3.3.1 for storage of Group A Plastic commodities at the storage and building height specified. The sprinkler system consists of Early Suppression

Table 1 Characteristic warehouse building sizes, commodities stored, sprinkler types and designs

Design	Commodity	Building Height (ft)	(m)	Sprinkler Temp. (°F)	(°C)	Sprinkler Response	Delivered Water Density (gpm/ft²)	(mm/min)	In-rack Sprinklers	Height of Sprinkler Above Floor (ft)	(m)	Height of Sprinkler above Storage (ft)	(m)
W1	Class II	15	4.6	165	74	Ordinary	0.17	6.9	No	15	4.6	3	0.9
W2	Plastic	15	4.6	165	74	Ordinary	0.6	24.4	No	15	4.6	3	0.9
W3	Class II	40	12.2	165	74	Quick	1.42	57.9	No	40	12.2	5	1.5
W4	Plastic	40	12.2	165	74	Quick	1.59	64.8	No	40	12.2	5	1.5
W5	Class II	100	30.5	165	74	Ordinary	0.3	12.2	Yes	15[a]	4.6	10	3.0
W6	Plastic	100	30.5	165	74	Ordinary	0.45	18.3	Yes	15[a]	4.6	10	3.0

[a] The in-rack sprinklers in the 100 ft tall warehouse are located 15 ft above the floor, and would be the first sprinklers expected to respond

Fast Response (ESFR) sprinkler heads with quick response RTI of 50 m$^{1/2}$ s$^{1/2}$ (91 ft$^{1/2}$ s$^{1/2}$) at 74 °C (165 °F). This system will provide each sprinkler with a flow rate of 477 lpm (126 gpm). No in-rack sprinklers are provided due to the use of ESFR sprinklers.

5.1.5 Warehouse Design 5

Warehouse design 5 (W5) consists of a 30.5 m (100 ft) tall warehouse with 18 tiers of Class II storage, reaching up to a height of 27.4 m (90 ft) with an ASRS. An automatic sprinkler system has been installed at the ceiling in accordance with NFPA 13 Table 16.3.1.2 for storage of Class II un-encapsulated commodities at 30.5 m (100 ft) storage heights. The sprinkler system was designed as a Control Mode/Density Area (CMDA) with ordinary temperature response at 74 °C (165 °F) and an RTI of 80 m$^{1/2}$ s$^{1/2}$ (145 ft$^{1/2}$ s$^{1/2}$). Horizontal barriers are installed within the racks at 4.6 m (15 ft) spacing and an in-rack sprinkler system installed in accordance with NFPA 13 Fig. 16.3.1.2a with the same sprinklers and design as the overhead is installed beneath each barrier with a staggered installation and 3.0 × 3.0 m (10 × 10 ft) spacing.

5.1.6 Warehouse Design 6

Warehouse design 6 (W6) consists of a 30.5 m (100 ft) tall warehouse with 18 tiers of Group A Plastic storage, reaching up to a height of 27.4 m (90 ft) with an AS/RS. An automatic sprinkler system has been installed at the ceiling in accordance with NFPA 13 Table 17.3.1.1 for storage of Group A commodities at 30.5 m (100 ft) storage heights. The sprinkler system was designed as a Control Mode/Density Area (CMDA) with ordinary temperature response at 74 °C (165 °F) and an RTI of 80 m$^{1/2}$ s$^{1/2}$ (145 ft$^{1/2}$ s$^{1/2}$). Horizontal barriers are installed within the racks at 4.6 m (15 ft) spacing and an in-rack sprinkler system installed in accordance with NFPA 13 Fig. 17.3.1.2a with the same sprinkler heads and design as the overhead is installed beneath each barrier with a staggered installation and 3.0 × 3.0 m (10 × 10 ft) spacing.

5.2 Characteristic Warehouse Fires

5.2.1 Fire Location and Ignition Scenarios

The warehouse designs have been evaluated with regard to several typical warehouse fire scenarios. Determination of typical warehouse fire scenarios is based on knowledge of real fires occurring in warehouse properties. In the United States, statistical fire data is compiled by the National Fire Incident Reporting System (NFIRS).

Such data is compiled from fire incident reports submitted by responding fire departments throughout the country. This data identifies the type of buildings involved in fires, the factors leading up to ignitions, the detection, suppression, fire fighting operations, as well as extinguishment procedures, and the losses of life and property resulting from such fires.

According to the statistical analysis, U.S. fire departments responded to an average of 1,350 warehouse structure fires annually during 2003–2006. An annual average of 21 civilian injuries, 5 civilian deaths, and $124 million in property damages have resulted from such fires [20]. These numbers do not include incidents occurring in cold storage properties, but according to the United States Energy Information Administration (EIA) statistics, these properties only represent about 2 % of all warehouse storage facilities, and thus were not considered for this analysis [4]. It is generally recognized that life safety/egress is not the primary fire problem in warehouse fires, but rather that the costs of property damages are considerably greater than for other structure fires [20]. The fire incidents considered for this analysis include those that cause the greatest amount of property damages.

The area of the building where the fire begins is of key interest to the development and growth of the fire. The location impacts the surrounding amount and type of materials, and thus the ability of the fire to spread and grow. Recent statistical analysis conducted in conjunction with this project by Everts reviewed warehouse fire data from 2005 to 2009 [21]. This work was developed as an update to existing warehouse fire statistical analysis conducted by Ahrens using NFIRS data compiled from 2003 to 2006 [19]. Factors contributing to property damages were determined by calculating the contribution of each factor toward the total amount of monetary losses resulting from warehouse fire incidents.

The size of a warehouse has been shown to impact both the likelihood and costs of fires. According to the EIA statistics only 15 % of warehouse properties are larger than 1,858 m^2 (20,000 ft^2) [4]. NFIRS data collected by Everts has shown that these properties account for 26 % of all reported warehouse fire incidents and 63 % of the total property damages [21]. The average cost of all reported warehouse fires from 2006 to 2009 was $123 K, but for warehouses over 1,858 m^2 (20,000 ft^2) the average cost was $297 K. The disproportionate number of fires and property damages from warehouses of this size demonstrate the need to consider additional risks in such properties. The design warehouses in this analysis have been placed into this category, with a size of 9,290 m^2 (100,000 ft^2).

The area of origin having the greatest contribution to property damages include storage areas (34 %), shipping/receiving or loading docks (13 %), processing or manufacturing areas (8 %), and storage rooms, tanks and bins (7 %). The fires represent 62 % of all warehouse fire property damages, and the remaining fires occur in various locations with small statistical contributions, including exterior walls, structural areas, offices, or roofs [21]. While the storage area consists of the greatest amount of combustible materials available for fire growth, the majority of fires are not initiated there. Consideration must be given toward fires developing elsewhere in the building. Such fires can become large and destructive even when isolated, but the potential to spread to the high rack storage areas would often result in the largest and costliest fire events.

In addition to the area of the building where the fire begins, the exact location of the fire within those spaces can also impact the potential for fire spread and growth, and the ability of the fire department to operate effectively. The fire growth rate is dependent upon the amount and proximity of combustible materials to the ignition point. Within high rack storage, fire growth rates are dependent upon the ignition location. Different fire growths would be expected for a fire occurring within the flue spaces between storage and one occurring within a boxed commodity. Similarly, different fire growth rates would be expected between a fire occurring at ground level or one occurring toward the top of the storage array. For this analysis, warehouse fires have been considered that are ignited within the high rack storage area and in the shipping and receiving area, where the potential for fire spread to the high rack storage exists. In addition, fires ignited within the high rack storage consist of a fire started in the flue space at the floor level and one ignited on top of the highest level of the rack storage.

The leading causes of fires contributing to total property damages in warehouses have been reported by Everts to be intentional (10 %), electrical distribution and lighting (10 %), heating equipment (6 %) [21]. These factors represent only about one-quarter of all fires; however all remaining fire cause classifications account for less than 6 % each including vehicles, smoking materials, torches, burners or soldering irons, and chemical reactions. The leading sources of heat identified in warehouse fires are sparks, embers, or flames from operating equipment (18 %), electrical arcing (17 %), a manual lighter flame (13 %), or unclassified heat from other powered equipment (8 %). These heat sources represent 46 % of the total property damages, and additional sources with smaller contributions include spontaneous combustion and chemical sources, molten or hot materials, smoking materials, friction, lightning and others. In order to represent the leading causes of ignition, fire scenarios investigated will include an intentional ignition, and ignition resulting from heating equipment, and an ignition caused by a malfunction in the electrical lighting system.

The types of materials first ignited in fires causing warehouse property damages include cardboard (11 %), sawn wood (9 %), fabrics and fibers (9 %), plastic (8 %), flammable or combustible liquids including gasoline (7 %). Numerous other materials have been identified with smaller contributions, including wood, particle board, paper, and rubber [21]. Evaluated fire scenarios have been developed considering ignitions occurring on external cardboard surfaces of palletized storage commodities. The selection of other materials would impact the ignition characteristics, the duration of the incipient phase, the fire growth rate, and the thermal output of the design fires. Each of these variables are examined by varying the cardboard ignition scenarios, and the use of only one material has greatly simplified the scope of the analysis.

Several fire scenarios have been developed with consideration to the statistical contributions of the materials first ignited, the source of the heat and ignition, and the location of the fires. The fire scenarios are summarized in Table 2. In addition

Table 2 Warehouse fire incidents

Fire scenario	Description	Scenario details		Potential fire growth			
		Ignition area	Potential first material ignited	Potential ignition cause	Incipient period	Growth rate	Max size
F1	Intentional ignition	High rack, floor level in flue	External cardboard box, gasoline	Intentional	Virtually none	Extremely rapid	Extremely large
F2	Top of rack ignition	High rack, top level, top surface	Cardboard box	Exploding light bulb	Highly variable	Variable, potentially rapid	Extremely large
F3	Shipping and receiving ignition with spread to high rack	Shipping and Receiving, adjacent to high rack	Non-racked pallets of the stored commodity, spreads to high rack	Radiated heat from operating equipment	Highly variable	Variable, depending on geometric configuration and combustible materials	Extremely large
F4	Shipping and receiving ignition without spread to high rack	Shipping and receiving, remote from high rack	Non-racked pallets of the stored commodity, unable to spread to rack storage	Radiated heat from operating equipment	Highly variable	Variable depending on geometric configuration and combustible materials	Limited to available fuel in area
F5	Long incipient period from smoldering or overheat	Electrical equipment	Wire insulation	Electrical overheat/ arcing	Long, overheat fire condition with delayed transition to flaming	Variable	Variable

to addressing statistically common fire scenarios, the assessed fires are intended to span a range of incipient fire durations, fire growth rates, and maximum potential fire sizes expected in warehouse properties. The potential impact of detection technologies for improving a manual fire department response are dependent upon the amount of time available until the fire size and the internal environment reaches a level hazardous enough that aggressive fire fighting operations cannot be safely or effectively conducted. The factors that impact that available time include the incipient duration, the fire growth rate, the amount of adjacent combustible material, the presence and effectiveness of sprinkler systems, the amount of smoke production, and the overall response time of the fire department.

5.2.2 Fire Development

With consideration to the effectiveness of a detection system, the primary differences between the fire scenarios described in Table 2 involve the incipient period duration, the fire growth rate, and potential peak fire size. With regard to these scenarios, the incipient period is considered as the time before flames begin to rapidly spread through combustible materials (i.e., before exponential fire growth). The incipient period may consist of hot embers, smoldering materials, or even flames too small to spread rapidly to adjacent combustibles. However, this period can also include small flaming fires of tens of kilowatts that grow slowly. An intentionally ignited fire in the flue space at the base of rack storage (F1) would most likely develop into an extremely large fire quickly, while an overheated electrical cable (F5) may take days or hours to develop into a hazardous fire.

The type of commodities stored within the racks directly impacts the fire growth rate and the potential peak HRR. The warehouse designs considered and shown in Table 1 include the use of either a Class II or Group A plastic commodity. These materials are described in the following section.

Commodities

A Class II commodity is one that consists of a noncombustible product in slatted wooden crates, solid wood boxes, multiple-layered corrugated cartons, or equivalent combustible packaging material, with or without pallets as defined by NFPA 13, Sect. 5.6.3.2 [3]. For this analysis, the Class II product consists of the FMRC standard Class II metal lined, double triwall cartons. Only the cartons contribute to fire growth. The cartons consist of corrugated cardboard 0.5 m (21 in) in each dimension and stacked $2 \times 2 \times 2$ on each pallet. Each double row rack contains 16 boxes, or 2 total pallets of material. A pallet of the FMRC Class II commodity has been characterized and its contents are shown in Table 3 [40].

The plastic commodity consists of a product that contains Group A plastics with corrugated cardboard packaging that may be stored with or without pallets. For this analysis, the plastic commodity is the Factory Mutual standard plastic test

Table 3 Composition of each pallet (Tier) of stored commodities considered for this analysis [38, Table 5.12]

Commodity	NFPA Class	Cardboard/ Paper [kg (lbs)]	Wood [kg (lbs)]	Plastic [kg (lbs)]	Total [kg (lbs)]	Combustion Energy [MJ]
Metal lined double triwall	Class II	36 (79)	26 (57)	0 (0)	85 (287)	1,100
Polystyrene Cups	Group A plastic	19 (42)	26 (57)	31 (68)	76 (168)	2,070

commodity. This consists of 0.45 kg (16 oz) crystalline polystyrene jars packaged in compartmented, single-wall corrugated cardboard boxes. The boxes measure approximately 0.5 m (21 in.) on each side and contain 5 layers of 25 polystyrene jars. Eight cartons, arranged $2 \times 2 \times 2$ are placed on a $1.1 \times 1.1 \times 0.1$ m ($42 \times 42 \times 5$ in.) slatted hardwood pallet. The contents of the commodity have been characterized for composition and heat content [40] and are shown in Table 3.

Fire Growth

Estimations of the free burning rates of the commodities without suppression or fire department intervention have been developed for the various fire scenarios. The fire sizes are based upon commodity fire test data that has been extrapolated and simplified to apply to the full range of fires, commodities, and storage configurations applied in this analysis. Although the estimated fire sizes are reliant upon several assumptions and extrapolations, they are intended to provide a rough picture of the development and extent of warehouse fires.

Full scale fire testing was conducted with the Class II and plastic commodities by Lee [41] and Spaulding [42]. This work determined the commodity fire growth rate coefficients and peak heat release rates for 8 total pallet loads, each consisting of $2 \times 2 \times 2$ box arrays, in 2 tiers of double row racks stored 2 deep beneath the FMRC fire products collector [38, Table 5.13]. Eq. (1) [38, 5.4.10] (known as the t^2 fire growth model) has been shown to be applicable for determining the free burn heat release rate (HRR) of racked commodities by Delichatsios [43]. The HRR for the Class II and plastic commodities in this analysis can then be determined by the fire growth rate coefficients, α, for the Class II and plastic commodities shown in Table 4. The incipient fire time, t_0, is the time required for the fire to reach a point

Table 4 Burning characteristics of a 2×2 arrays of single tier storage commodities determined by Lee and Spaulding for commodity free burn growth rates in 2 Tier double row racks and scaled linearly for a single tier [38, Table 5.13]

Commodity	α (MW/s^2)	Q_{max} (MW/tier)	Q_{total} (MJ/tier)
Class II	0.000105	4.05	4,400
Plastic	0.00045	11.25	8,280

where the flame growth can occur, either by a smoldering transition to flame, a change in the available oxygen, modified geometric conditions such as entering a flue space, or through various other mechanisms of fire development. The time required to reach 60 % of the peak HRR can be calculated by Eq. (2).

$$Q = \alpha(t - t_0)^2 \quad Q < 0.6Q_{max} \tag{1}$$

where

Q = heat release rate (kW)
α = commodity fire growth rate coefficient in kW/s^2
t = time in seconds
t_0 = incipient fire time in seconds

$$t_{60\%} = t_0 + \sqrt{0.6Q_{max}/\alpha} \tag{2}$$

Q_{max} = The peak HRR obtained for the commodity
Q_{total} = The total amount of thermal energy contained in the commodity.

The storage unit used to measure the free burn growth rates, consisting of 2 tiers of a double deep, double row rack is shown in Fig. 3. The burning rate of Class II and plastic commodities has been shown to increase linearly with the number of tiers of storage for 2, 3, and 4 tier storage racks [38, Figs. 5.7 and 5.8]. Without test data conducted on the scale of the 7 and 18 tier storage racks considered for the 12.2 m (40 ft) and 30.5 m (100 ft) warehouse designs considered for this analysis, the linear relationship is assumed to extrapolate. Thus, the burning rate of the rack storage for any number of tiers, N, has been estimated to be calculated by Eq. (3), and the time to reach 60 % of the peak HRR is obtained by Eq. (4).

The total exposed surface area of a single tier of double deep, double row storage, 2 × 2, is 27.3 m^2 (294 ft^2), summing contributions from all exposed sides.

Fig. 3 Single modeled combustible unit for double row rack storage consisting of 8 pallets stored in 2 tiers with dimensions in feet

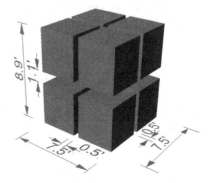

At the peak burning rate, or when the pallet is fully involved, the HRR per unit area, Q'', for the Class II and plastic commodities is calculated to be 0.15 MW/m^2 and 0.41 MW/m^2, respectively. The exposed surface area is directly proportional to the number of tiers of the storage racks and the lateral extent of the flame spread, and thus the total HRR will also be proportional to the height and floor area of the fire spread. The exposed surface area per floor area for each tier of the storage is 5.25 $m^2_{surface}/m^2_{base}$.

$$Q = N\alpha(t - t_0)^2 \quad Q < 0.6\,Q_{max} \tag{3}$$

$$t_{60\%} = t_0 + \sqrt{0.6Q_{max}/\alpha} \tag{4}$$

When the HRR has developed beyond 60 % of the peak value, the t^2 fire growth presented in Eq. (3) has been observed to become linear as the flame spreads to the edges of the boxes [40]. The overall growth of the fires were shown to decrease, as the limits of the fuel were reached, but such edge effects are not of concern when the fire is ignited in the high rack area, as the fire will continually spread to surrounding commodities.

A linear growth rate is estimated due to lateral flame spread along the flues at a rate of approximately 2 cm/s (0.78 in./s). This value has been estimated from rack storage thermocouple data obtained by Hamins and McGrattan [44]. The total growth in the fire HRR due to the lateral flame spread rate is calculated by determining the additional total exposed surface area resulting from the 2 cm (0.78 in.) lateral spread.

As the fire continues to spread laterally, the combustible materials behind the flame front are consumed and eventually exhausted. The total fire size will continue to grow linearly due to flame spread until the first ignited material is consumed. At some time, the reduction in fire size due to the exhaustion of fuel will be equivalent to the increase in fire size due to lateral flame spread, and the fire size has been assumed to reach a steady state (ignoring the eventual spread across aisles).

The amount of available fuel energy, ΔE, contained in each 0.02 m (0.78 in) incremental segment can be determined by multiplying by the width of the row, 2.29 m (7.5 ft), the number of tiers, N, and the heat content of the double row per tier, Q_{total}, and dividing by the base area of the double deep, double rack storage, 5.2 m^2 (28 ft^2) as shown in Eq. (5).

$$\Delta E[MJ] = 0.02(2.29)NQ_{total}/5.2 = 0.0088NQ_{total} \tag{5}$$

The HRR of each 0.02 m (0.78 in) segment, ΔQ, can be determined by multiplying by the number of tiers, N, the width of the double row racks, 2.29 m (7.5 ft), the HRR per unit exposed surface area area, Q'', and the exposed surface area per base area, 5.25 m^2/m^2 (ft^2/ft^2), as shown in Eq. (6).

$$\Delta Q[MW] = 0.02(2.29)(5.25)NQ'' = 0.240NQ'' \tag{6}$$

Dividing the amount of available energy contained in each segment by the HRR of each segment, yields the amount of time until all the fuel is consumed. The time would be the time that the 0.02 m (0.78 in) segments would begin to extinguish each second, at the same rate additional segments are igniting, thus a steady state HRR is obtained. The time to reach steady state HRR is calculated in Eq. (7).

$$t_{ss} - t_{60\%} = \frac{\Delta E}{\Delta Q} = 0.0367 \frac{Q_{total}}{Q''} \tag{7}$$

The steady state HRR achieved can then be calculated by adding the t^2 peak HRR and the HRR of from the lateral flame spread at time t_{ss}. The flame is estimated to spread in both directions along the storage row, and thus two 0.2 m (0.78 in) elements are added to the fire size each second. The steady state HRR is shown in Eq. (8).

$$Q(t_{ss}) = 0.6NQ_{max} + 2\Delta Q(t_{ss} - t_{60\%}) = N(0.6Q_{max} + 0.018Q_{total}) \tag{8}$$

The total HRR of the rack storage fires can then be calculated by Eq. (9) for any time until the fire has spread beyond the area of the warehouse or suppression is applied. The HRR per number of tiers for the Class II and plastic commodities as a function of time after the incipient period are shown in Fig. 4. The steady state fires would be expected to continue until the fire had spread to additional racks, increasing the overall fire size, or until the fire had spread to the limits of the building and the fire size was reduced by limiting fuel, or until oxygen depletion in the space limited and reduced the combustion rate.

$$\begin{array}{ll} Q \approx 0 & t < t_0 \\ Q = N\alpha(t - t_0)^2 & t_0 < t < t_{60\%} \\ Q = 0.6NQ_{max} + 0.48NQ''(t - t_{60\%}) & t_{60\%} < t < t_{ss} \\ Q = 0.6NQ_{max} + 0.018NQ_{total} & t > t_{ss} \end{array} \tag{9}$$

Testing was conducted by Hamins and McGrattan using the standard plastic commodities ignited by flaming propane burners [45]. When exposed to flames on the cardboard surface, an initial burning of the cardboard was followed by a flame

Fig. 4 Estimated heat release rate per tier of double row rack storage after incipient fire period ends and rapid flame growth begins. Spread across aisles is not included in this analysis

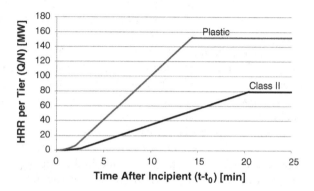

shortening as the fire moving into the burning commodities. A period of approximately 20–40 s was observed before the fire growth period began. This flaming exposure is similar to that used for the intentional ignition scenario for this test, with an incipient period prior to accelerate fire growth in the range of 20–40 s. For this analysis, the moment this fire is ignited, rapid flame spread begins in the racks and the fire develops as shown in Fig. 4, with the total HRR determined by multiplying by the number of tiers of storage.

When the fire is ignited by hot glass shards being deposited on the top of the rack (F2), the incipient period prior to the development of flue space fire spread is dependent upon the temperature of the hot glass, the initial burning location on the top of the box, and various environmental and geometric parameters. The incipient time of this fire could range anywhere from minutes to hours, depending upon whether the cardboard is subject to flaming or smoldering ignition. Limited research has been done in characterizing incipient fire growth times. Much of this work has focused on cigarette ignitions [e.g., Holleyhead], Holleyhead [46] reports that no flaming was observed for lighted cigarettes placed on horizontal cardboard, but against vertical arranged sheets of cardboard, ignition occurred after 15 min. The potential for higher heat capacity in hot glass shards would be expected to lead to a greater propensity for ignition than cigarettes.

After the end of the incipient phase when the fire transitions to the rapid growth phase, the fire would be expected to develop and spread similar to the fire in Fig. 4. When the fire is ignited on the top of the boxes, however, the fires would not be able to spread vertically upward, and have been modeled as fires occurring in single tier racks and any potential downward fire spread has been neglected. As such, the fire growth could be larger than modeled.

When the fire is ignited on pallets of the stored commodities in the shipping and receiving area (F3, F4), the incipient time may also vary between seconds and hours. Once the fire is initiated and rapid fire growth can begin, the burning pallets have been estimated to burn in the same manner as for the single tier of storage shown in Fig. 4, limited by the extent of the pallet array in the shipping area. For the development of a numerical analysis, the amount of material in this space has been estimated to be 4 pallets of material, stacked 1 pallet high in a 2 × 2 array. This fire reaches a steady state HRR when all the material has been involved. The fire is ignited along one face of the pallets and begins to grow after the incipient stage according to the t^2 model of Eq. (3). As this fire grows beyond the valid t^2 region, above 60 % of the peak HRR, the fire growth begins to become affected by the boundaries of the fuel and reduces to an approximate linear rate until reaching the peak. Testing conducted by FM for 2, 3, and 4 tier storage racks confirm this linear fire growth [40]. Utilizing the data from these tests, a relationship for the linear growth rate, α_{linear}, can be estimated as a function of the t^2 growth rate coefficient, α. Calculation of this coefficient yields α_{linear} equal to 33.3 $\alpha_{Class\ II}$, and α_{linear} equal to 37.8 $\alpha_{Plastic}$. Application of these linear fire growth rates has been applied until the 2 × 2 array has reached the estimated peak HRR. The estimated free burn HRR of the 4 pallets located in the shipping and receiving area after the incipient period for the Class II and plastic commodities is shown in Fig. 5.

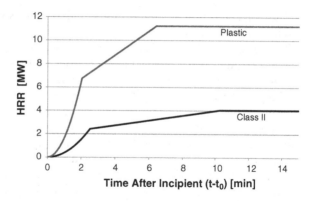

Fig. 5 Estimated HRR after incipient period of 4 pallets in 2 × 2 array ignited in the shipping and receiving area

In addition to the unknown incipient period, the ability of this fire to spread beyond the shipping and receiving area and the amount of time required to ignite the main high rack storage is extremely variable. The ignition of the rack depends upon the development and transport of sparks or embers, and the radiated heat flux to the storage. Both of these depend upon the proximity of the fire to the rack storage. If this fire is capable of spreading from this area to the high rack storage the overall size, potential for destruction, and hazard to occupants and fire fighters can significantly increase.

When the incipient period of a fire is very long, such as for the overheated electrical wire scenario (F5), the fire maintains a limited HRR with great potential for detection before it transitions to a more significant hazard.

The general findings of this fire growth analysis include:

- The fire sizes presented in this portion of the analysis do not consider the impact of automatic suppression systems or manual response.
- Peak fire sizes obtained for plastic commodity fires are generally 2 times that obtained for Class II commodities. The fire growth rates of these fires are also greater, growing at nearly 4 times the rate. The amount of time to provide a response to fires involving plastic commodities is substantially reduced compared to the storage of Class II commodities.
- Fire sizes calculated for ignitions in the high rack storage do not account for "aisle jumping" and thus actual fires may be larger than those estimated for this analysis.
- The fire sizes obtained for ignitions located in the flue space at the base of storage racks are capable of becoming extremely large (80–150 MW) fires within a few minutes if not contained.
- Fires ignited at the top of storage may have significant incipient periods, but can develop into large fires (80–150 MW). These fires were estimated as occurring in only 1-tier of storage and downward flame spread was not considered. Therefore, actual fires may be larger as drop down of burning materials will involve lower tiers.

- The incipient period of the intentional ignition (F1) is very short (on the order of a minute) and the overheated wire (F5) can be very long (on the order of hours). The incipient period of the ignition of materials from deposition of a hot exploded light bulb (F2) or radiation from overheated equipment (F3, F4) is highly variable and may range from seconds to hours.
- The peak HRR of the shipping and receiving fire is limited by the amount of available material. This analysis has considered this material to be a 2 × 2 array of four pallets of the stored commodity. Larger or smaller quantities would increase or decrease this peak HRR accordingly.
- Due to the use of general correlations and broad assumptions in the development of the estimated HRR, results have been rounded and presented as a general range of minutes of development and MW of fire size when used to review the impact of detection. The numbers presented in Figs. 4 and 5 merely provide general guidance about the range of expected fire outcomes.

Smoke Layer Development

Smoke production during fire events is a fundamental component of the fire characterization. While the HRR remains low during the incipient phase, it increases rapidly during the escalating fire growth phase. Production of smoke during the incipient phase provides a potentially detectable indicator of the presence of the fire while smoke production during later stages and during suppression and extinguishment may act to limit visibility and the overall ability to fight fires.

Limited analysis was performed on the design fires using Consolidated Model of Fire and Smoke Transport (CFAST) [47] to provide an estimate of when the hot smoke layer within the warehouse fire would reduce visibility and limit fire department operations. For fires occurring in the high rack areas, the fire sizes are proportional to the number of tiers, the height of the building, and thus the total building volume. Consequently, smoke filling times are comparable for each rack fire scenario in the different building heights. The time required for the smoke layer depth to drop below 1.5 m (5 ft) is on the order of 10 min. Given the large fire sizes at this time, both visibility and extent of fire would prohibit effective manual response, regardless of detection. With fires starting in other areas away from the rack storage, smoke filling times will be longer and may be sufficient to allow fire department access to the seat of the fire. In these cases, fire detection can potentially have an impact.

These estimated smoke filling times were obtained for warehouses without heat and smoke vents. If installed, the design of such devices would utilize similar fire hazard estimations and develop smoke removal systems intended to match the estimated smoke production curves in accordance with NFPA 204 [48]. If these devices were properly designed to match the fire hazards within the warehouse, the time required to fill the building with smoke and reduce visibility could be significantly extended. Thus, detection has even more potential impact to affect fire department response with smoke and heat vents.

When the sprinkler systems operate and contain the fires, the water droplets will cool the smoke and drag the smoke downward. This pushes the smoke layer down, and can further reduce visibility at the floor. In addition, even though the fire growth may be reduced or eliminated by the sprinkler system, the production of smoke may continue. Consequently, fire detection that can provide fire department response before sprinkler activation is a key performance goal.

Sprinkler Activation

The fire growth estimates described in Sect. 5.2.2 are applicable to free burning commodities without active suppression systems. If a suppression system has been properly designed and installed and is operational, the fire sizes should be controlled after sprinkler operation. Utilizing the fire plume temperature and velocity correlations presented by Heskestad and Delichatsios [49] and the heat transfer solution by Beyler [50], the activation times of sprinklers can be modeled for fires with t^2 growth rates. The correlations take into account the lag time for the plume to reach and spread across the ceiling, the growth of the flame, and the response time index (RTI) of the detection element. The system of equations is solved to determine the lag time for hot gases to reach the sprinkler head location, t_{lag}, and the time required for a given head to reach the alarm threshold temperature.

The temperature at a sprinkler above the fire and an estimated time for sprinkler activation for each fire scenario is presented below. The activation time is a function of the fire growth rate, the physical location of the sprinkler with respect to the fire, and the RTI of the sprinkler. The height of the sprinklers above the fires has been estimated as the distance from the top of storage to the sprinkler height.

Sprinklers for these tests scenarios are 74 °C (165 °F), ordinary temperature with either fast or ordinary response times, having an RTI of 50 $m^{1/2}$ $s^{1/2}$ (91 $ft^{1/2}$ $s^{1/2}$) or 80 $m^{1/2}$ $s^{1/2}$ (145 $ft^{1/2}$ $s^{1/2}$), respectively. All sprinklers are spaced 3.1 × 3.1 m (10 × 10 ft) on the ceiling and within racks, and thus a radial distance of 2.2 m (7.1 ft) has been used as the maximum distance from the fire to the sprinkler. The activation of in-rack sprinklers beneath horizontal barriers has been calculated in the same manner as for ceiling sprinklers, except the ceiling height has been replaced with the height of the horizontal barrier. The activation of in-rack sprinkler calculation provides a rough estimate, because there is not a free plume and ceiling jet as exists with normal ceilings used to develop the correlations. The calculated sprinkler activation times and the fire HRR at activation for each of the fire scenarios in each design warehouse are shown in Table 5.

The general sprinkler activation analysis has shown that:

- Most fires in the storage racks activate sprinklers quickly within 1 min and at HRR of 1–2 MW.
- When the fires are ignited in the shipping and receiving area, they are located a maximum distance beneath the sprinkler heads. When the ceiling is extremely high, as for the 30.5 m (100 ft) ceilings, activation is not achieved for the design

Table 5 Sprinkler activation times and HRR at activation for each fire scenario in each warehouse design

Warehouse	Commodity	Fire scenario	Height of sprinkler above top of storage (m, ft)	Sprinkler activation after incipient (min)	HRR at time of sprinkler activation (MW)
W1	Class II	Intentional	0.9 (3.0)	0.9	0.6
4.6 m		Top Rack	0.9 (3.0)	1.1	0.4
(15 ft)		Ship/Rec	3.5 (11.5)	1.5	0.8
		Incipient	NA		
W2	Plastic	Intentional	0.9 (3.0)	0.6	1.0
4.6 m		Top Rack	0.9 (3.0)	0.7	0.8
(15 ft)		Ship/Rec	3.5 (11.5)	0.9	1.4
		Incipient	NA		
W3	Class II	Intentional	1.5 (5)	0.6	0.8
12.2 m		Top Rack	1.5 (5)	1.0	0.3
(40 ft)		Ship/Rec	11.1 (36.5)	2.5	2.3
		Incipient	NA		
W4	Plastic	Intentional	1.5 (5)	0.4	1.5
12.2 m		Top Rack	1.5 (5)	0.7	0.7
(40 ft)		Ship/Rec	11.1 (36.5)	1.5	3.5
		Incipient	NA		
W5	Class II	Intentional	0.3 (1.1)[a]	0.8	0.5
30.5 m		Top Rack	3.1 (10)	1.4	0.7
(100 ft)		Ship/Rec	29.4 (96.5)	$T_{max} = 36\ °C\ (96\ °F)$[b]	
		Incipient	NA		
W6	Plastic	Intentional	0.3 (1.1)[a]	0.6	1.0
30.5 m		Top Rack	3.1 (10)	0.9	1.3
(100 ft)		Ship/Rec	29.4 (96.5)	$T_{max} = 50\ °C\ (122\ °F)$[b]	
		Incipient	NA		

[a] Height from storage to in-rack sprinkler located below horizontal barrier
[b] HRR required to achieve sprinkler activation is not reached during fire scenario and no activation is expected, peak sprinkler temperature from steady state fire is calculated by Alpert's correlation for steady state fire sprinkler activation [51]

fires. If this fire is sustained for a significant period, ceiling sprinklers will be preheated to an estimated 35–50 °C (95–122 °F). If the fire is able to spread to the high racks or to other available combustibles, a large number of sprinkler heads may activate. This can result in reduced flow per head and a failure to control fires. This represents a particularly hazardous potential fire scenario.

- Design fires ignited in the shipping and receiving area under the 12.2 m (40 ft) ceiling are able to activate sprinklers, but not until fire sizes of 2–4 MW are achieved.
- The fires ignited under the 12.2 m (40 ft) ceiling involve ceiling-only ESFR sprinklers. The water delivery of the ESFR is intended to provide enhanced suppression over the control mode sprinklers used in the other warehouse designs and this may help further reduce the size of the fires compared to the control mode design used in the other warehouses.

Impact of Suppression on Fire Growth

Testing has been conducted to attempt to quantify the impact of sprinklers on rack storage fires. It has been suggested by Yu et al. that the fire sizes will decay exponentially as a function of the delivered water density if the applied density is sufficient [52]. It has also been suggested by Hamins and McGrattan that the fires will initially decay exponentially followed by a slow linear fire growth [45]. The ability of the sprinkler system to reduce the HRR of the fires, control fire spread and growth, or accomplish extinguishment depend upon the delivered water density, the ability of the spray to penetrate the hot fire plume and reach the burning material. To provide conservatism for this analysis, it is assumed that at the time of sprinkler activation, the spread of flames to additional material will stop and the fire will be controlled. The fire will continue to burn in its existing locations at the HRR at the time of first sprinkler operation as presented in Table 5. Burning will continue until the available fuel is consumed or the fire department has reached the seat of the fire and performed extinguishment operations and overhaul. If the fire department cannot access the seat of the fire, the fire may grow out of control.

Due to the cooling impact of the sprinklers on the fire plume and the downward thrust of the sprinkler spray, the smoke layer is dragged down to floor level and expected to fill the space surrounding the fire. The expected visibility within the space surrounding the fire after the activation of sprinkler systems is assumed to be limited.

5.2.3 Fire Department Response

The previous sections of the fire analysis were directed at determining the potential growth rates, maximum free burn fire sizes, and expected sprinkler response characteristics of the design fires and warehouses. In conjunction with the characteristics of the fires themselves, the response of the fire department is a fundamental element when considering the potential hazards of such warehouse fires. The interior conditions of the space, the size and extent of the fire, and the amount of smoke all impact the actions taken by the fire department to protect the warehouse property. The response time of the fire department plays a key role in determining the status and development of these conditions. A general outline of the elements that comprise the fire department response time, developed by the United States Fire Administration (USFA), is shown in Fig. 6 [53].

In general, the dangers associated with fighting fires in warehouses are increased by modern warehouse design characteristics, including larger buildings, larger storage areas, and impediments related to ASRS. These hazards can be reduced if the fire department is able to arrive when the fire is smaller in size and the amount of heat and smoke is reduced. Earlier notification of incipient fire events could help the fire department to reach fires earlier in the growth period and increase available firefighting time.

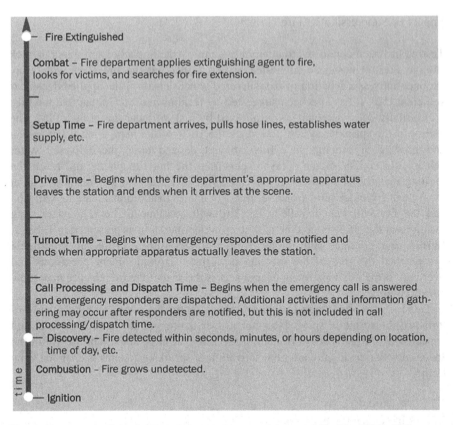

- Fire Extinguished

Combat – Fire department applies extinguishing agent to fire, looks for victims, and searches for fire extension.

Setup Time – Fire department arrives, pulls hose lines, establishes water supply, etc.

Drive Time – Begins when the fire department's appropriate apparatus leaves the station and ends when it arrives at the scene.

Turnout Time – Begins when emergency responders are notified and ends when appropriate apparatus actually leaves the station.

Call Processing and Dispatch Time – Begins when the emergency call is answered and emergency responders are dispatched. Additional activities and information gathering may occur after responders are notified, but this is not included in call processing/dispatch time.
- Discovery – Fire detected within seconds, minutes, or hours depending on location, time of day, etc.

Combustion – Fire grows undetected.

- Ignition

time

Fig. 6 Components of the fire department total response time as defined by USFA [53]

The fire department response time calculated in national statistics consists of the period following notification until the first apparatus arrives on the scene. Nationally, response times have been shown to be less than 5 min in nearly 50 % of incidents, less than 8 min in 75 % of incidents, and less than 11 min in 90 % of incidents [53]. The 5–10 min response times account for the time to transfer and interpret the alarm, the time to dress and ready, and the travel time from the department to the scene. These times do not account for the amount of time after arrival required to setup and plan activities. Including setup time, a reasonable estimate of the time required from initial notification until combat begins is 10–15 min.

When arriving at fire incidents, the primary goal of the fire department is to save lives within the building. Most often when dealing with warehouse fires, however, the occupants have escaped and are not in danger. The primary losses of warehouse fires are associated with property damages. According to NFPA 1500, the firefighter rules of engagement after evaluating the scene consider the following [54]:

- "We will risk our lives a lot, in a calculated manner, to save savable lives."
- "We will risk our lives a little, in a calculated manner, to save savable property."
- "We will not risk our lives at all for a building or lives that are already lost."

In general, if the fire department believes that they can save the building while putting themselves in limited risk, then they will attempt to protect the property. Evaluation of the fire scene is often based upon available fire pre-planning, including knowledge of the warehouse storage configuration, commodities, and existing fire protection systems. Even when supplied with adequate information about the scene, there have been numerous reports of fire chiefs adopting a tactic of "if everyone is out, we are not going in" when dealing with warehouse fire events. There is often an unrealistic expectation about the tactics to be used by the fire department to protect the property when designing fire protection systems for warehouse properties [55].

When considering whether to enter a burning warehouse property, the fire department is generally concerned with several elements of the event, including existing knowledge of the facility, the visibility within the space, the overall size of the property, and the access routes available to them, and knowledge of the actual fire location within the building. Knowledge of the facility is developed during preplanning for fire events. In addition to the property characteristics, the preplan normally includes the firefighting strategy and the tactics to be employed. The fire department response and operations may be impact by the availability of trained and equipped plant personnel to assist with operations [25].

Visibility within the facility is a key aspect required for effective firefighting. In general, the density of smoke within the space affects the fire department's ability to locate the seat of a fire, and to determine safe travel routes in and out of the facility. Often, fire departments will work to improve visibility through venting. Opening up external walls to provide horizontal ventilation as well as roof venting are often used to reduce the amount of smoke. While manually shutting down sprinklers can help increase visibility, it has been shown to contribute to fire control and losses [26]. Shutting off sprinklers in this manner is explicitly prohibited by NFPA 13E, Sect. 4.3.3 [27].

In addition to visibility, the size of the building plays a key role in providing access for tactical firefighting operations. A large floor area may result in a large travel distance required to reach the fire location, requiring long hose lays and increased air bottle consumption. Available air bottle supplies and duration may not be adequate to allow for safe entry and exit [55].

The height of the warehouse also plays a key role in firefighting. In addition to allowing increased vertical fire spread and larger overall fires, the vertical location of a fire can make firefighting nearly impossible. A firefighter standing in a 1.2 m (4 ft) wide aisle trying to apply water to a fire located 12.2 m (40 ft) above would require a nearly vertical hose spray, and successful application of water to the burning fuel would prove impossible. The hazards presented by fires in tall storage result in high risks to firefighters and reduced likelihood of an interior attack [26].

5.3 Warehouse Fire Analysis

5.3.1 Intentional Ignition Scenario (F1)

Fire scenario F1 consists of an intentional ignition of a small volume of accelerant poured in a flue space in the center of the storage array at the floor level. There is an extremely short expected incipient period and rapid fire growth is expected to begin at ignition. Expected HRRs for this fire scenario have been calculated for warehouses of 4.6 m (15 ft), 12.2 m (40 ft), and 30.5 m (100 ft) tall heights containing either Class II or Group A plastic commodities. The HRRs have been estimated both with and without operational sprinkler systems providing containment and stopping the fire growth and flame spread. Approximate sprinkler activation times after ignition and the HRR at that time for the intentional ignition scenarios are shown in Table 6.

Sprinklers operate quickly for all intentional fire scenarios. The rate of vertical and lateral fire spread within the double row racks results in sprinkler activations within approximately 1 min or less for the Class II and plastic commodities. The peak fire sizes reached for 30.5 m (100 ft warehouse are of comparable magnitude to the 4.6 m (15 ft) warehouses because the 30.5 m (100 ft) height warehouses have in-rack sprinklers at 4.6 m (15 ft) vertical spacing, which is the same as ceiling level sprinklers in the 4.6 m (15 ft) warehouses. The fire sizes at activation in the 12.2 m (40 ft) warehouses are slightly greater than for the other heights, reaching 1–2 MW. This is due to the use of ESFR sprinklers at the ceiling only with no in-rack. The ESFR sprinklers are designed to deliver additional water

Table 6 Approximate sprinkler activation times and HRR at activation for intentional ignition (F1) fire scenario

	Commodity	Warehouse design height	Sprinkler present and effective		
			Height of sprinkler above top of storage (m,ft)	Sprinkler activation after end of incipient stage (min)	HRR at time of sprinkler activation (MW)
Intentional ignition in flue space at base of high rack area	Class II	W1 (15 ft)	0.9 (3.0)	1	<1
		W3 (40 ft)	1.5 (5)	<1	1
		W5 (100 ft)	0.3 (1.1)[a]	<1	<1
	Plastic	W2 (15 ft)	0.9 (3.0)	<1	1
		W4 (40 ft)	1.5 (5)	1	1–2
		W6 (100 ft)	0.3 (1.1)[a]	<1	1

[a] Height from storage to in-rack sprinkler located below horizontal barrier

density with larger droplet sizes to provide enhanced suppression over the control mode sprinklers used in the other designs. Thus, the intensity of these fires should be reduced by the sprinklers more than with standard sprinklers.

The potential impact of detection is limited for these scenarios. The sprinkler systems activate quickly, within approximately 1 min, and the expected incipient time is between 20–40 s [45]. Detection systems would have marginal effect even if used to activate suppression systems and would not be able to provide additional time for firefighting operations.

If there is no sprinkler system activation, the intentional fire scenario is capable of growing out of control. HRRs greater than 100 MW can occur, even for the Class II commodity, with only 2-tiers of storage in the 4.6 m (15 ft) warehouse within 12–13 min of ignition. It is very unlikely that any entry or aggressive internal firefighting tactics would be used, and a total loss would be expected. The warehouses with taller storage or plastic commodities are expected to reach even greater HRRs in less time, and thus interior fire fighting would also not be expected.

The non-sprinklered intentional ignition fires reach uncontrollable levels in less than 10 min, which is the earliest projected time the fire department would begin fire fighting operations. With limited or no incipient period before fire growth, even if detection were able to identify the fire at the exact instant of ignition, the fire department would still arrive at a fire they would not actively fight. If there were on-site trained personnel, fire detection may afford a benefit to control the fire, particularly if backed up with off-site fire department assistance.

5.3.2 Exploding Light Bulb Igniting the Top Rack of Storage (F2)

Fire scenario 2 (F2) consists of a fire ignited by an exploding light bulb depositing hot glass onto the top of a cardboard box in the top rack of storage. The incipient time duration required for this scenario to develop into a flaming fire and the subsequent growth of the fire may vary significantly. The box may ignite quickly or may smolder for several minutes or even hours prior to flaming fire growth. In addition, the flames may be slow to spread across the top of the box and into the surrounding commodities. However, once significant fire growth begins, these fires have the potential to spread throughout the stored commodities and develop into extremely large fire events beyond the capability of arriving fire departments in the absence of operable sprinklers. Expected HRRs for this fire scenario have been determined for warehouses of 4.6 m (15 ft), 12.2 m (40 ft), and 30.5 m (100 ft) heights containing either Class II or Group A plastic commodities after the variable incipient time. The HRRs have been estimated both with and without an operational sprinkler system providing fire control. Approximate sprinkler activation times after the start of the fire growth phase and the HRR at that time are shown in Table 7.

The top rack of storage is located close below the ceiling sprinkler system, and this allowable distance is limited for various storage heights by NFPA 13 [3].

Table 7 Approximate sprinkler activation times and HRR at activation for top rack storage ignition scenario (F2)

	Commodity	Warehouse design height	Sprinkler present and effective		
			Height of sprinkler above base of fire (m, ft)	Sprinkler activation after end of incipient stage (min)	HRR at time of sprinkler activation (MW)
Intentional ignition in flue space at base of high rack area	Class II	W1 (15 ft)	0.9 (3)	1	<1
		W3 (40 ft)	1.5 (5)	1	<1
		W5 (100 ft)	3.1 (10)	1–2	<1
	Plastic	W2 (15 ft)	0.9 (3)	<1	<1
		W4 (40 ft)	1.5 (5)	<1	<1
		W6 (100 ft)	3.1 (10)	<1	1–2

When a fire ignites in this location, once fire growth begins the sprinkler systems are able to operate quickly when the fire is less than 1 MW. After activation, the sprinkler system has relatively unobstructed access to the origin of the fire. However, if the fire starts to drop down or is partly shielded below the top tier, the fire could continue, but should be controlled.

The greatest potential impact of detection systems in this scenario would depend upon the duration of the incipient fire period. If the incipient period was long enough to allow early warning detection and a manual response prior to fire growth and sprinkler activation, significant property damage could be prevented. In this case, the damage prevented would include that done by the fire as well as water damage. If the fire could not be detected to allow fire department operations prior to sprinkler operation, the impact of detection would be limited.

If the sprinkler system does not operate properly, the top rack fire scenario is capable of growing out of control. Despite a conservative calculation that neglects fire spread across aisles or downward from the top tier of storage, HRRs greater than 20 MW were estimated even for the Class II commodity at all storage heights with 5–7 min. It is very unlikely that successful entry and aggressive internal firefighting operations would be possible for such a fire, resulting in a total loss. Even if the fire department could enter the space, there would be limited ability to fight a fire occurring on the top racks of storage in taller facilities.

If provided with detection during the incipient phase, the fire department could potentially take actions prior to the initiation of fire growth. While the fire is still incipient, fire fighters could enter the space and use the lift equipment in the warehouse to reach the seat of the fire at the top of the rack and perform extinguishment and overhaul. After fire growth has begun, such actions would not occur, and little could be done to fight the fire from the floor levels for storage 12.2 m (40 ft) or higher.

The non-sprinklered top rack ignition fires reach uncontrollable levels within a few minutes. Even with notification at the initiation of the accelerated fire growth period, the fire department would be unable to begin fire fighting operations. The potential impact of detection for this scenario would depend entirely upon the ability of detection to operate during the incipient period and provide enough time for a manual response before escalating fire growth. Given the estimated total fire department response time of 10–15 min, any detection device that could identify the incipient fire and notify the fire department more than 15 min before exponential fire growth would be effective.

5.3.3 Ignition in the Shipping and Receiving Area (F3, F4)

Fire scenario 3 and 4 (F3, F4) consist of the ignition of non-piled pallets of commodities in a shipping and receiving area of the warehouse. For this analysis, the ceiling heights in the shipping and receiving area are considered equivalent to the height in the storage area. This is a worst case design as ceiling heights in actual warehouses may have reduced heights in the non-rack storage areas.

Although this fire may initiate in numerous ways, the designed cause of ignition for this scenario is considered to be radiation from operating equipment in close proximity. After an indeterminate incipient period, fire growth begins in the pallets in the shipping and receiving area. The potential for the fire to remain contained to combustibles in the shipping and receiving area (F3) or spread to the high rack storage area (F4) are considered.

The design fire for this scenario consists of a total of 4 pallets of single tiered commodity in the shipping and receiving area. The total amount or configuration of this material can significantly impact the fire growth rate, the achieved peak HRR, and the ability of the fire to spread to the high rack storage area. The sprinkler activation times after the incipient period and the estimated HRR at activation are shown in Table 8.

Table 8 Approximate sprinkler activation times and HRR at activation for shipping and receiving fire ignition scenario (F3, F4)

	Commodity	Warehouse design height	Sprinkler present and effective		
			Height of sprinkler above top of storage (m, ft)	Sprinkler activation after end of incipient stage (min)	HRR at time of sprinkler activation (MW)
Intentional ignition in flue space at base of high rack area	Class II	W1 (15 ft)	3.5 (11.5)	1–2	<1
		W3 (40 ft)	11.1 (36.5)	2–3	2–3
		W5 (100 ft)	29.4 (96.5)	$T_{max} = 36\ °C$ (96 °F)[a]	
	Plastic	W2 (15 ft)	3.5 (11.5)	<1	1
		W4 (40 ft)	11.1 (36.5)	1–2	3–4
		W6 (100 ft)	29.4 (96.5)	$T_{max} = 50\ °C$ (122 °F)[a]	

[a] HRR required to achieve sprinkler activation is not reached during fire scenario and no activation is expected, peak sprinkler temperature from steady state fire is calculated by Alpert's correlation for steady state fire sprinkler activation [51]

When the fires are ignited in the shipping and receiving area, they are located a significant distance below the ceiling sprinkler heads. For the 4.6 m (15 ft) and 12.2 m (40 ft) warehouses the activation times occur generally within 1–3 min and with fire sizes of 1–4 MW. When the height of the warehouse is further increased to 30.5 m (100 ft), however, the sprinklers are not capable of activated for the fire sizes provided by the 2 × 2 array of pallets. It is estimated by Alpert's correlation for steady state fires [51] that a minimum fire size of 27 MW would be required to activate sprinklers in this scenario.

When activation does not occur, the ambient ceiling temperatures are increased and the fire is allowed to continue to burn. If the fire is able to spread to adjacent combustibles, including the rack storage, numerous sprinkler heads may activate due to the elevated ambient ceiling temperatures. When too many sprinkler heads activate, the pressure is significantly reduced and the delivered water density decreases. This could result in ineffective sprinkler operation and an inability to control the fire. Early warning detection of the shipping and receiving fire could provide time for the fire department to respond before the fire can spread. This could prevent a situation where a failed sprinkler system resulted in a total building loss.

As was the case with the ignition at the top rack of storage (F2), the impact of detection is also dependent upon the incipient time before fire growth. If a detection system could provide notification longer than 10–15 min before fire growth begins, the initiation of the fire growth and the activation of the sprinkler system could potentially be prevented. Even after fire growth, there are a few minutes when early detection notification could improve the response time over the sprinklers and may allow the fire department to prevent some damage or prevent a potential spread to the high rack area.

If the sprinkler system does not operate as intended, the shipping and receiving fire scenario is generally limited by the amount and configuration of combustibles in the area. If the fire is contained to this area, the non-sprinklered fire sizes estimated for the 4 pallet load reach peak fire sizes of approximately 4 or 11 MW for the Class II and plastic commodities, respectively (see Fig. 4). While these are large fires, the location of the shipping and receiving are is likely near an access point and the lack of rack storage may improve the ability of the fire department to locate and extinguish these fires. Even if the sprinkler system does not operate, depending upon the combustible loading in the area, detection may be able to provide sufficient notification at any point during the incipient period or fire growth. Besides direct thermal fire damage, early detection will also limit smoke damage throughout a building, which could be very costly depending on the stored commodities.

5.3.4 Overheated Electrical Cable (F5)

Fire scenario 5 (F5) consists of a fire initiated by an overheated electrical wire first involving wire insulation. This scenario represents a fire with an extremely long incipient period. During this period smoke is produced but limited heat and flame. This type of fire does have the risk of developing into a large fire, but may take minutes, hours, or even days to develop to such a state. Fire threats such as these are ideal cases for detection equipment that can quickly identify the smoldering condition and provide notification for a manual response prior to development into a hazardous condition.

5.4 Additional Applications of Detection Equipment

In addition to improving the response time of fire departments arriving at warehouse fire events, early warning detection can also provide additional benefits. It is possible that detection equipment can provide information to the fire department during setup and assessment, such as the internal temperatures in the space, the depth and visibility of the smoke layer, the location and extent of the fire spread, or even visual images of the conditions inside the space. This information can be extremely valuable when the fire department is making an assessment of how, or whether, entry should be undertaken. If more information is available, they may determine that safe entry and fire fighting could be conducted in incidents where otherwise a successful interior attack might not be undertaken.

Although this analysis has focused on the response and intervention of off-site fire departments, it is recognized that some facilities may have trained, on-site personnel that can provide more rapid response to fire detection alarms. Obviously, fire detection can have a larger impact on a broader range of fire scenarios with appropriate on-site response. It is anticipated that on-site response could be achieved within 3 min. In some cases, the introduction of fire detection may facilitate and justify the need for on-site response whereas without detection, the conditions would be deemed to hazardous for on-site personnel. Incipient fires can often be approached and extinguished through the actions of trained employees, including removing the electrical power supplies, removing adjacent combustibles, and even use of portable extinguishers.

As noted for multiple scenarios, including both off-site and on-site manual fire responses, fire detection is particularly advantageous when fires can be detected sufficiently before the fire transitions to an accelerated growth phase. Unfortunately, defining clear performance requirements for the different fire scenarios is not possible since incipient growth stages are not well defined, particularly for palletized and rack stored commodities. In order to provide additional context for this variability, several examples from the available literature are discussed. Ignition of upholstered furniture has been examined for the duration of incipient fire development for various ignition locations and sources by Collier and Whiting at the Building Research Association of New Zealand (BRANZ) [56]. They investigated the use of cigarette brands, matches, and a small gas flame. Ignition with the cigarette brand was sporadic, but a flaming fire was achieved with an incipient phase of 12 min prior to t^2 fire growth. Testing with flaming sources yielded an incipient phase ranging from 88–130 s. Additional testing of incipient duration on upholstered furniture by Mitler and Tu [57] with a small propane burner indicated incipient growth periods ranging from 10–36 min. Other work analyzed by Babrauskas for smoldering to flaming upholstered furniture items show a range of transition times from 22 to 306 min [58]. Improving the understanding of incipient growth phases for more typical warehouse commodities in palletized or rack storage configurations would improve the assessment of fire detection and provide a foundation for specific performance requirements.

Another potential use of early detection technology could be initiation of advance suppression and extinguishment systems. The use of heat detection for such a purpose has already been included in NFPA 13 for use with high expansion foam systems in warehouse environments [3]. Additional work was performed investigating various advanced warehouse extinguishment options as part of the High Challenge Warehouse Workshop [2]. It is possible for future applications that early detection in warehouses could be used for initiation of advanced suppression and extinguishment systems.

5.5 Fire Hazard Analysis Summary and Conclusions

Several basic warehouse designs and realistic warehouse fire scenarios were evaluated to assess the potential impact of fire detection to allow fire departments to respond at earlier stages of a fire. Based on typical off-site fire department response and setup times to attack a fire, it was assumed that the fire service would require 10–15 min to combat a warehouse fire. Therefore, fire scenarios were assessed as to whether detection systems could realistically provide early enough warning to affect a significant difference in fire damage relative to the 10–15 min fire department response.

Typical commodities, including cardboard and plastic, that are ignited by established fires (e.g., arson) were shown to have fast fire growth rates within rack storage, reaching upwards of 100 MW in size (with no suppression systems) before fire departments would be able to respond. Smoke layers also descended to near the floor about the same time, reducing visibility and further hindering manual operations. Consequently, without required overhead and in-rack sprinkler systems, fire detection provides little benefit for reducing damage for fires in rack storage once they transition to an accelerated growth rate.

Even with a functioning sprinkler system, fire detection will not have a substantial impact on most fire scenarios occurring in the high rack storage unless the fire is detected in the incipient phase where personnel can respond before sprinkler activation. This minimal impact is because the sprinkler systems will respond in most cases within several minutes limiting fires to less than 5 MW. Due to the relatively fast response of the suppression system well before a manned response can occur, detection has minimal effect.

However, for non-rack storage locations, such as shipping and receiving, in tall warehouses (i.e., over 40 ft), detection can have an effective impact even with functioning sprinkler systems. This is because the high ceiling sprinkler systems cannot respond until fires become quite large (e.g., >20 MW for a 100 ft high building), subsequently presenting a significant fire challenge and also posing a risk to rack involvement. When sprinkler activation does not occur, the ambient ceiling temperatures are increased and the fire is allowed to continue to burn. If the fire is able to spread to adjacent combustibles, including the rack storage, numerous sprinkler heads may activate due to the elevated ambient ceiling temperatures. When too many sprinkler heads activate, the pressure is significantly reduced and the delivered water density decreases. This could result in ineffective sprinkler operation and an inability to control the fire.

As with all scenarios, the potential impact of a fire detection system is directly related to the duration of the incipient stage before accelerated fire growth. Regardless of suppression system activation or location of the fire, if detection is able to provide a warning 10–15 min before the end of the incipient stage, the fire department should be able to limit damage. In the event that there is on-site, trained personnel for fire response, fire detection will have a larger impact on a broader range of fire scenarios, assuming a response within 3 min. Therefore, with on-site trained personnel, fire detection would need to provide warning more than 3 min before the end of the incipient stage.

Longer incipient stages can be characteristic of smoldering to flaming ignition sources, such as a ruptured overhead light fixture onto packaging, or improperly discarded smoking material. Pyrolizing materials due to overheating components, such as electrical wires, can also be a source of long incipient stages. Providing specific time periods is difficult as this is an area in need of much work. There is limited data on incipient fire growth times. This is particularly true for palletized commodities and rack storage configurations.

6 Research Plan

The fire hazard analysis has demonstrated the potential impact of detection technologies to reduce property damages from warehouse fires. The impact depends upon the duration of the incipient fire period and the ability of the technology to detect incipient conditions. There are many scenarios where warehouse fire detection can be an integral part of limiting fire damage and the interruption of business operations. In particular, a detection technology capable of detecting a fire before it has reached a rapid development stage presents the greatest benefit when it allows an effective manual response. The fire science literature provides limited insights into the two key elements of assessing the benefit of warehouse detection (incipient growth times and detector responses to realistic warehouse scenarios).

Fire modeling of warehouse fire growth from ignition through the incipient phase and the rapid growth phase is beyond the current state-of-the-art.

Consequently, real-scale fire testing is needed to define warehouse fire events and to determine the performance of various detection technologies to these representative fire events. This experimental approach will concentrate on characterizing the development of initiating warehouse fires and assessing the performance of detection technologies to these fires including consideration of warehouse related obstructions. The phase 2 work outlined below will form the basis for assessing other issues, such as nuisance source rejection and situational awareness (i.e., fire location and growth information). However, before these issues can be addressed, the design fires must be established and the baseline detection performance must be determined. For example, the required sensitivity or installation criteria of a detection system to effectively detect the design fires must be established before nuisance immunity is evaluated. In addition, since certain nuisance sources are only applicable to specific technologies, evaluation of nuisance sources is a secondary evaluation to the phase 2 work discussed below.

The incipient stage fires will include commodity packaging (e.g., the outside exposed material) or other miscellaneous materials for fires that initially occur away from the rack storage. As discussed in Sect. 4.2.1, the material first ignited causing the most warehouse property damages were cardboard and fabrics. The ignition of warehouse fires was most commonly the result of electrical distribution (14 %) and intentional arson (13 %). However, the arson fires contributed most significantly toward the damage of property, and are the cause of 21 % of all warehouse fire damages [19]. The complete breakdown of the causes of warehouse structure fires determined by NFIRS and NFPA Surveys is shown in Fig. 1. Other leading causes include heating equipment and a category called torch, burner or soldering iron. It is apparent that ignition sources of interest include both flaming fires and a range of other heat sources. Obviously, arson fires that are created with a substantial initiating fire will have no appreciable incipient stage. Therefore, only small flaming fires, as might be created with a cigarette lighter or arcing/flaming electrical fault, are of interest.

The phase 2 work will include two main objectives:

1. Characterize the incipient fire growth phase of representative warehouse fire scenarios to establish standard design fires for evaluating detection system performance.
2. Evaluate detection system performance for a range of design fires and representative warehouse conditions.

The characterization of design fires can be accomplished with real-scale laboratory tests with standard fire test instrumentation. The evaluation of detection system performance would also be conducted at full-scale with minimum facility parameters of 20–30 ft height and detector-to-fire distances of 75 ft and greater. The incipient fire growth evaluation would focus on cardboard packaging (single and double wall corrugated with the use of interior foam packaging behind cardboard would be another fuel package arrangement). Other materials associated with fires initiating away from rack storage would also include additional materials such as fabrics (e.g., rags) and electrical cables.

Ignition of the source material would be accomplished with a range of heat sources and a small pilot flame. The heat sources would replicate scenarios ranging from a cigarette, to hot glass (i.e., from an exploding overhead light), to a hot forklift exhaust manifold or heater next to the material. Some of these heat sources can be used directly or can be simulated via standard heating sources (e.g., nichrome wire and a range of cartridge heaters or radiant panels).

The development of a fire is also dependent on the configuration of the source material and the ignition source. Consequently, source materials will be systematically evaluated under different geometrical configurations. In addition to geometrical configurations, material edge effects can also play a role. For example, a particular heat source exposing the side of a piece of fabric or cardboard box may have limited effect, but if the same source is exposed to the frayed edge of the fabric or edge of the cardboard, a fire may be initiated. Available air flow conditions will also impact the ignition and development of the fire sources.

Three primary geometrical configurations will be evaluated, particularly for the cardboard: flat surface, parallel surfaces, and an interior corner configuration. The more material is brought together in a parallel fashion, the more heat will increase due to the proximate surfaces. Therefore, the three configurations represent the bounding and intermediate configurations possible relative to having no re-radiation between surfaces (single flat surface) to ideal radiation between parallel surfaces of material.

Fuel scenarios evaluated will consist of ignition exposures to both single material items and also composite sources (i.e., a cardboard box with another combustible material inside). It is expected that some ignition source/material combinations may not develop for the single flat surface configuration. However for boxes with a flat surface exposure, the fire may propagate through the material to the interior (either empty or with a combustible, such as foam packaging) and continue to grow. Without a combustible interior, such as with a Class II commodity, the flat surface exposure may self-extinguish. Flat surface exposures could occur in a horizontal scenario, such as a hot glass fragment falling on the top of a box, or in a vertical scenario, such as a heater of hot manifold up against the side of a box.

Although very limited, the existing technical literature on incipient fire development will be evaluated and used to guide efforts in phase 2. For example, work by Holleyhead summarizes ignition of solid materials by cigarettes [59]. The work describes the potential for cigarettes to ignite cardboard relative to a few configuration effects. Babrauskas provides a review of several studies that examined radiant flux criteria for the ignition of cardboard as well as a study of cardboard ignition from fire brands of different sizes and wind conditions [58]. Based on the literature and the phase 2 testing, ignition source propensity to cause a warehouse fire will be documented relative to the size/intensity of the source.

Characterization of the tested fire scenarios will include measurements of the times of development and transient measurements of the heat release rate, the smoke development, and the radiant heat output. The time of development would be established from the initiation of the event to the time the fire transitions to flaming, the time to selected fire sizes, and the time to when the fire transitions to a rapid growth phase. These times will form the bases for whether detection systems can be effective as discussed in the fire hazard analysis above (e.g., respond in sufficient time to allow a manual response).It is anticipated that the incipient stages will be well below 100 kW, but it is recommended that fires be allowed to progress to tens of kilowatts to establish the rapid growth phase transition. Based on the gasoline soaked ignition sources used to evaluate sprinkler activation to rack storage fires, 50–100 kW is deemed to be the size at which palletized cardboard box fires will accelerate quickly.

The heat release rate (HRR) is the principle measurement that characterizes the size of the fire. The HRR provides a means for identifying the incipient and rapid growth rate stages of a fire. Establishing the HRR is useful for developing standards with good test to test repeatability and for providing a benchmark for assessing detection system sensitivity. Measuring smoke development and radiant heat output (only applicable for flaming stages) also provide key descriptive parameters that relate to the principles of detection for a number of systems. The measured values for the different fire scenarios will serve to establish a database of characterized design fires that can be used in future standards or as inputs to fire models, which would be a significant improvement in the state-of-the-art of modeling detection systems.

The evaluation of potential warehouse fire scenarios will be used to identify specific scenarios that would be the basis for detection system performance. For example, fire scenarios that have sufficient incipient stages (i.e., greater than 3 min relative to on-site fire response teams or 10 min relative to local fire departments) would be deemed as representative of fires for which detection systems would be effective. Therefore, such fire scenarios would serve as design fires for evaluating detection system performance.

Once design fires are established, the second objective of the phase 2 work would be to evaluate detection system performance for a range of fires. Different detection technologies would be evaluated simultaneously while being exposed to the design fires. The first step would be to assess detection performance with fires directly in the open (relative to line of sight or overhead sensors). A second series of tests would include a range of general obstructions as may occur for fires remote from the rack storage. A third series of tests would include an array of obstructions related to fires in and adjacent to racks. The response times for various fire scenarios would be a function of the test compartment size and height and environmental conditions. Initial testing would be conducted for a single geometric building size, but further evaluation into the impact of building size and environmental conditions (i.e., ventilation and temperature) could be investigated.

Fire detection system response would be assessed based on the ability of a system to detect the range of design fires (with and without obstructions) and, secondly, the time of response relative to providing sufficient manual response times. Sufficient manual response times will be characterized using two criteria of 3 and 10 min before accelerated fire growth. The results of the detection testing will assist in the development of evaluation standards for warehouse fire detection and will also provide a basis for evaluating the effectiveness of state-of-the-art detection systems in warehouse environments. Ultimately, effectiveness will need to be assessed relative to a set of commodity damage criteria, which can range from direct fire damage to smoke damage.

References

1. Fire Protection Research Foundation. (2009). Storage fixed fire protection and final extinguishment, *Core Planning Meeting*, Sheraton Braintree, MA, January 2009.
2. Gallagher, R., & Gollner, M. (2010). High challenge warehouse case study—summary, *Warehouse workshop*, NFPA Fire Protection Research Foundation, Orlando, FL, February 2010.
3. NFPA 13. (2010). *Standard for the installation of sprinkler systems*, 2010 Edition, Quincy: National Fire Protection Association.
4. United States Energy Information Administration (EIA). (1999). Building Characteristics by Activity: Warehouse and Storage, 1999.
5. Gallagher, R. (2011). Warehouse challenge. *NFPA Journal, 105*(4), 42.
6. NFPA 230. (2003). *Standard for the fire protection of storage.* Quincy: National Fire Protection Association.
7. International Fire Code (IFC). (2009). *Chapter 23—High piled combustible storage,* International Code Council.
8. Allen, T., & Molina, R. (2006). Warehouse sprinkler design configurations not covered by NFPA 13. *Fire Protection Engineering, 29,* 16–18.
9. Golinveaux, J., & Hankins, J. (2006). Meeting the challenges of an ever-changing storage industry. *Fire Protection Engineering, 29,* 32–40.
10. Building Research Establishment. (2006). *Sprinkler installation trends and fire statistics for warehouse buildings.* Department for Communities and Local Government.
11. Parlor, B. (2003). Managing the fire threat in warehouse and distribution buildings. *Fire, 96*(1173), 28–29.
12. Fire Precautions (Workplace) Regulations, 1997.
13. Regulatory Reform (Fire Safety) Order, 2005.
14. Kidd, S. (2001). Fire protection in warehouses. UKWA Raising Standards, March 26, 2001.
15. CTIF Working Group. (2003). Firefighter's safety in large single storey storage buildings. International Technical Committee for the Prevention and Extinction of Fire (CTIF), 2003.
16. Khan, N. (2010). Special Report. Middle East Warehouse Fires, Arabian Supply Chain.com, July 18, 2010, available: http://www.arabiansupplychain.com/article-4517-special-report-middle-east-warehouse-fires/.
17. Haq, R. (2011). Dubai Police Report 22% Rise in Warehouse Fires. Arabian Supply Chain.com, February 9, 2011, available: http://www.arabiansupplychain.com/article-5567-dubai-police-reports-22-rise-in-warehouse-fires/.
18. Goves, A. (2003). The Japanese approach to warehouse fire safety. *Fire, 96*(1172), 24–25.
19. Ahrens, M. (2009). Warehouse fires excluding cold storage. National Fire Protection Association, February 2009.
20. Hall, J. (2003). Warehouse information. National Fire Protection Association, February 2003.
21. Everts, B. (2011). Structure fires in warehouses. National Fire Protection Association, September 2011.
22. Hall, J. (2011). U.S. Experience with Sprinklers. National Fire Protection Association, May 2011.
23. Fire Statistics. (2007). Department of Communities and Local Government, United Kingdom, 2007.

24. De Smet, E. (1996). General Warehouse in New Orleans, Louisiana on March 21, 1996, Fire Risk Assessment Method for Engineering (FRAME) case study, 1996.
25. Frank, J. (2010). Will the fire department Intervene. *Industrial Fire World*, 25(1).
26. Nadile, L. (2009). The problem with big. *NFPA Journal*, March/April 2009, pp. 46–49.
27. NFPA 13E. (2010). *Recommended practice for fire department operations in properties protected by Sprinkler and Standpipe systems*. Quincy: National Fire Protection Association.
28. Tremblay, K. (2010). Firewatch. *NFPA Journal*, *32*, July/August 2010.
29. Tremblay, K. (2010). Firewatch. *NFPA Journal*, *26*, January/February 2010.
30. Tremblay, K. (2011). Firewatch. *NFPA Journal*, *48*, May/June 2011.
31. Harrington, J. L. (2006). Lessons learned from warehouse fires. *Fire Protection Engineering*, *29*, 22–24.
32. NFPA 72 (2010). *National fire alarm and signaling code*, Quincy: National Fire Protection Association.
33. Gottuk, D. (2008). Video image detection systems installation performance criteria. Fire Protection Research Foundation, October 2008.
34. Gan, G. (2008). Warehouse Fire Detection Case Study. Fire Australia, Winter 2008.
35. Building Research Establishment. (2010). Smoke Detection in High Ceiling Spaces—Report 3. *BRE Global*, November 19, 2010.
36. Hadjisophocleous, G. et al. (2010). Study of video image fire detection systems for protection of large industrial applications and atria. Carleton University, InnoSys Industries, Inc. and General Fire Technologies, 2010.
37. Schirmer Engineering Corporation. (2010). Video image detection: Comparative testing of various detection technologies. axonX, LLC, Feb 2010.
38. High Expansion Foam Protects New Compact Storage System. (2010). News from Fire Technology at SP Technical Research Institute of Sweden, *Brand Posten* Number 41, 2010.
39. Hume, B., & Eady, M. (2002). The use of CFD computer models for fire safety design in buildings: Large warehouse case study. Office of the Deputy Prime Minister, Fire Research Division, 2002.
40. Zalosh, R. G. (2003). *Industrial fire protection engineering*. West Sussex: Wiley.
41. Lee, J. L. (1987). Stored commodity fire test program part 1: Fire products collector tests, prepared for the Society of the Plastics Industry, Factory Mutual Research Corporation, July 1987.
42. Spaulding, R. D. (1988). Evaluation of Polyethylene and Polyethylene Terephthalate commodities using the fire products collector. Factory Mutual Research Corporation, May 1988.
43. Delichatsios, M. A. (1983). A scientific analysis of stored plastic fire tests. *Fire Science and Technology*, *3*, 73–103.
44. Hamins, A., McGrattan, K., & Stroup, D. (1998). International Fire Sprinkler, Smoke and Heat Vent, Draft Curtain Fire Test Project: Large scale experiments and model development. Technical Report, National Fire Protection Research Foundation, Quincy, MA, 1998.
45. Hamins, A., & McGrattan, K. (2003). Reduced scale experiments on the water suppression of a rack-storage commodity fire for calibration of a CFD fire model. *Proceedings of the Seventh International Symposium on Fire Safety Science*, 2003, pp. 457–468.
46. Holleyhead, R. (1999). Ignition of solid materials and furniture by lighted cigarettes. A review. *Science & Justice, 39*(2), 75–102.
47. Peacock, R. D., Forney, G. P., Reneke, P., Portier, R., & Jones, W. W. (1993). *CFAST, The consolidated model of fire growth and smoke transport, NISTIR 1299*. Gaitherburg: National Institute of Standards and Technology.
48. NFPA 204. (2007). *Standard for smoke and heat venting*. Quincy: National Fire Protection Association.
49. Heskestad, G., & Delichatsios, M.A. (1978). The initial convective flow in fire, 17th Symposium on Combustion, Combustion Institute, Pittsburgh, PA, 1978.
50. Beyler, C. (1984). A design method for flaming fire detection. *Fire Technology, 20*(4), 9–16.

51. Alpert, R. (1972). Calculation of response time of ceiling-mounted fire detectors. *Fire Technology, 8*, 181–195.
52. Yu, H.Z., Lee, J.L., & Kung, H.C. (1994). Suppression of Rack Storage Fires by Water. *Proceedings of the Fourth International Symposium on Fire Safety Science*, 1994, pp. 901–912.
53. U.S. Fire Administration/National Fire Data Center. (2006). Structure fire response times. *Topical Fire Research Series, 5*(7), January 2006 (revised August 2006).
54. NFPA 1500. (2007). *Standard on fire department occupational health and safety program.* Quincy: National Fire Protection Association.
55. Frank, J. (2009). Intervene or leave it alone. *Industrial Fire World, 24*(6), November 2009.
56. Collier, P. C. R., & Whiting, P. N. (2008). Timeline for incipient fire development. Study Report No. 194, Building Research Association of New Zealand, Porirua City, New Zealand, 2008.
57. Mitler, H. E., & Tu, K. M. (1994). *Effect of ignition location on HRR of burning upholstered furniture, NISTIR 5499.* Gaithersburg: National Institute of Standards and Technology.
58. Babrauskas, V. (2003). *Ignition handbook.* Issaquah: Fire Science Publishers.
59. Holleyhead, R. (1999). Ignition of solid materials and furniture by lighted cigarettes. A review. *Science & Justice, 39*(2), 75–102.

Bibliography of Additional Sources

Ahrens, M. (2000). *Fires in plastic and plastic product storage facilities*. National Fire Protection Association.

Atkinson, G. T., & Jagger, S. F. (1993). Assessment of hazards from warehouse fires involving toxic materials. *Interflam*, 1993.

Bruce, J. (2007). *Preventing and managing ESFR protected warehouse distribution center fires in rialto*. Rialto Fire Department

Duval, R. (2000). *Storage Warehouse, Phoenix Arizona*. Fire Investigations, National Fire Protection Association.

Fang J. et al. (2009). An experimental evaluation about multiple fire detectors in a high large volume space. *14th International Conference on Automatic Fire Detection*, 2009.

Gollner, M. J., et al. (2011). Warehouse commodity classification from fundamental principles, part 1: Commodity and burning rates. *Fire Safety Journal, 46*, 305–316.

Gollner, M. J., et al. (2011). Upward flame spread over corrugated cardboard. *Combustion and Flame, 158*, 1404–1412.

Hall, J. (2003). *Warehouse information*. National Fire Protection Association.

High Challenge Warehouse Workshop. (2010). *Fire protection preliminary design concepts, SUPDET 2010*. The Fire Protection Research Foundation, Various Presenters.

Incident Recording System. (2009). *Questions and lists version 1.4*. Department for Communities and Local Government.

Liu, W. et al. (2010). Test research on air sampling-type smoke detection system operating in large open space. Institute of Building Fire Research, China Academy of Building Research (CABR).

National Standard GB50116-98. (1999). *Specifications for automatic fire alarm system design*. Beijing: China Planning Publishing House.

NFPA 92B. (2009). *Guide for smoke management systems in malls, atria, and large areas*. Quincy: National Fire Protection Association.

Oliver, T. (1988). Warehouse fires: A case for robotics. *Electrical Review, 221*(2), 26–27, 29.

Overholt, K. J., et al. (2011). Warehouse commodity classification from fundamental principles, part II: Flame heights and flame spread. *Fire Safety Journal, 46*, 317–329.

Peterson, K., & Markert, F., (eds.) (1999). Assessment of fires in chemical warehouses: An overview of the TOXFIRE project. Riso National Laboratory.

The Building Regulations. (2000). *Fire Safety – Approved Document B*. Office of the Deputy Prime Minister.

Zurich Services Corporation. (2009). Warehouse fire prevention—loss prevention. *Risktopics*.

J. Dinaburg and D. T. Gottuk, *Fire Detection in Warehouse Facilities*,
SpringerBriefs in Fire, DOI: 10.1007/978-1-4614-8115-7,
© Fire Protection Research Foundation 2012